Praise For
Breaking the Barriers to Belonging

With this book, the Losambes have penned an important contribution to the burgeoning discussion about belonging—what it means and the science behind it, why it matters, what interrupts it, and how we can harness its deep power to transform not just ourselves in relation to others, but the systems around us that desperately need to be re-imagined… with *all* of us in mind. They have issued a call to all of us, to invest in our shared humanity, in these pages. Let us heed it!

—Jennifer Brown, Founder & CEO, Jennifer Brown Consulting; Best-Selling Author, *Inclusion, How to be an Inclusive Leader,* and *Beyond Diversity* (with co-author Rohit Bhargava)

∗∗∗

Shared humanity! Is that not the quintessential goal of DEIB practitioners? This book is one-of-a-kind and a must-read for anyone who strives to balance emotion with evidence. Pascal and Crystal Losambe not only discuss the practical aspect of the human need to belong but also take the readers on a scientific discovery of how humans are neurologically hardwired to desire and seek belonging amongst one another. The authors describe the processes that occur in human brains when confronted with ambiguity, stress, negative experiences, and feelings of exclusion. These neurological responses are the same for everyone. This is our shared humanity.

—Dr. Nika White, President and CEO, Nika White Consulting, Author of *Inclusion Uncomplicated*

***

This powerful book explores belonging as a basic human need that is dictated by the basic workings of the human brain. Pascal and Crystal are able to boil down complex brain science into powerful wisdom that leaders, teachers, and really all of us need to create the spaces we want to create but don't know how. These are spaces where diverse individuals are seen and heard and able to step into the best, and most powerful, versions of themselves. Amazing contribution!

—Ross Wehner, Founder, World Leadership School, and Co-Founder of teachUNITED

***

The Losambes' book could not be any more timely than at the present time. The book also offers practical paths to deal with and resolve problems arising from discrimination, whether on an individual, collective, or institutional level. It is a book for everyone, a great source of information for totally unconnected individuals or for groups, because it foregrounds our common humanity or Ubuntu as an important collective frame of reference. It provides several ways to engage prejudicial harms for the sake of seeking individual solace and collective harmony, all towards social integration.

—Dr. Kasongo Kapanga, Professor of French and Chair of the Department of Languages, Literatures and Cultures, University of Richmond. Author of The Writing of the Nation: Expressing Identity through Congolese Literary Texts and Films

***

"Breaking the Barriers to Belonging" is an original! Pascal and Crystal Losambe weave together compelling stories, unexpected evidence, and fresh perspectives, delivering a valuable guide to navigating these incredibly divisive times. Each section offers actionable insights to help create moments of belonging. This book is a must-read for anyone looking to gain a better understanding of themselves and their ability to influence the culture around them.

—Rhodes Perry, CEO, Rhodes Perry Consulting and Bestselling Author of *Belonging at Work* and *Imagine Belonging*

Breaking the Barriers to Belonging

United by Our Biology and Shared Humanity

by

**Dr. Pascal Losambe,
Crystal Losambe**

Published by KHARIS PUBLISHING, an imprint of
KHARIS MEDIA LLC.

Copyright © 2023 Dr. Pascal Losambe & Crystal Losambe

ISBN-13: 978-1-63746-172-3

ISBN-10: 1-63746-172-0

Library of Congress Control Number: 2022946019

All KHARIS PUBLISHING products are available at special quantity
discounts for bulk purchases for sales promotions, premiums, fund-raising,
and educational needs. For details, contact:

Kharis Media LLC
Tel: 1-479-599-8657
support@kharispublishing.com
www.kharispublishing.com

Preface

"There is a reason the word 'belonging' has a synonym for 'want' at its center; it is the human condition."
—Jodi Picoult

Welcome. As you engage with this work, we sincerely hope you will be provided with knowledge, skills, and tools to bring about racial unity, reconciliation, and healing in your organizations and spheres of influence. We are the founders and leaders of Synergy Consulting Company, an organization designed to educate the masses on the human need to belong. Why? We have personal and professional knowledge about the impact of social discrimination. We have encountered barriers to belonging from our peers, leaders, and law enforcement that have resulted in us feeling devalued, unsupported, and disconnected. Consequently, we have experienced the impact of these situations on our own biases as well as personal and professional relationships. We've seen the negative impact of isolation, oppression, and low self-esteem in our lives, just being completely transparent and honest.

The greatest impact on our relationships was during the government elections and when reading the stories of minorities' deaths during encounters with law enforcement. When trying to explain our perspectives, concerns, and fears, we were often met with what seemed like a lack of empathy and understanding with no desire to view the situation from the other side. These same people may have even thought the same of us. Instead of productive, positive conversations, the encounters would turn into tense, emotional, and unproductive arguments. Offense was taken on both sides, which led to the dissolution of seemingly strong friendships. These same encounters occurred with people we knew, and consequently, we witnessed polarization within social groups and educational, religious, and professional organizations. Unfortunately, we are not the only people who have encountered such chaos and divisiveness.

As a society, we have not been taught how to wade through the difficult conversations and unique perspectives and come out the other side as friends. We do not believe this is how God intended human relationships to be. We are called to show compassion, empathy, grace, and mercy towards others. We are taught to love others with no exceptions, which is much easier said than done. Because we have lived through racial discrimination and felt the pain of losing meaningful relationships, we understand that racial healing will only happen with education, humility, perseverance, interdependence, and forgiveness.

Humans were made to belong in a community of people, not to live isolated and alone. It is in our biology. Our inquisitive nature and thirst for knowledge, coupled with our interests in many of the world's processes, moved us towards the fields of science and medicine. As we journeyed through school and our careers, we became fascinated with the brain and its impact on human behavior. The information we absorbed and devoured allowed Pascal to pursue a Bachelor's in Molecular Biology and Biochemistry, a Master's degree in Biology with a focus on Neurobiology, and a Ph.D. in Educational Leadership with a focus on Cultural Competence. Crystal, a Nurse Practitioner, has a Bachelor's in Healthcare Administration and Master's in Nursing with a focus on Family Medicine and Rheumatology, and she is currently a candidate for a Doctor of Nursing Practice degree.

Our study of science and medicine has allowed us to learn how the human brain operates and how we all have a physiological and psychological need to be a part of a whole, or a need to belong, if you will. We have made it our mission to share this information with the world because it is critical to healing the obvious division in modern-day society. The impetus to this book is traceable back to our youth and through our years of learning and experiences, both positive and negative. This book will explore why we act or feel in specific ways from a scientific perspective and how we can use that knowledge to make this world more inclusive for all through our shared humanity.

We believe we can be better when we know more and begin to heal once we understand we are all more alike than different, which is how we were originally designed. When we view life through the lens of our shared humanity, we will implore empathy and compassion for one another. We will break down barriers that separate us and move us to a more inclusive society where we can truly appreciate and celebrate our uniqueness AND commonalities.

Our Wish

Our mission is to help individuals and organizations become more empathetic and more aware of their words, beliefs, and actions. Why? Because no one should feel isolated or alone at home, with friends, or at work. We all possess the basic biological need to belong and be in relationship with others. Our hope is that this book, just like Synergy Consulting Company, will be a stepping stone to cultivating a heart of inclusivity, leading to profound shifts in organizational and societal cultures where no one feels ostracized because of their uniqueness.

We look forward to taking this journey to unity with you as we share life-altering information that will benefit everyone.

Contents

Introduction

"My humanity is caught up in yours."
—Archbishop Desmond Tutu

We live in a global, interconnected world marked by international commerce, communication, and migration. Because of this, understanding how to connect with people from various backgrounds is a necessity in our personal and professional lives. There is a lot of great discussion and literature around the topic of diversity, equity, and inclusion (DEI), but it is more confined to the academic environment and often ineffective when translated to real-world scenarios. Moreover, there is often a lack of understanding of humans' biological nature and its impact on DEI. Linking our common biological needs and the necessity for inclusion becomes imperative when discussing our shared humanity. The content of this book, therefore, bridges the academic discussion with practical behavior patterns and the real-world impact of these topics. We look at DEI topics from a neurobiological perspective and show how the brain influences our experiences and behaviors. We use evidence-based approaches when providing recommendations for racial reconciliation and inclusive practices. At the baseline of our existence as humans, we all want the same things and have the same needs. Looking at DEI from this perspective will provide a way for everyone to feel included in the discussion.

The book's purpose is to explain the biological need to belong and how the biology of the brain guides our perspective and relationship with others. Another purpose of the book is to enlighten readers about the shared humanity perspective and how it explains our need for community and relationship in a diverse group of people.

Dr. Pascal & Crystal Losambe

We wrote this book for individuals and leaders who yearn for unity, reconciliation, and racial healing in their organizations, our nation, and the world. This book is also for minorities and people of color who must endure the pains of bias, discrimination, and marginalization and who need to put their emotions into words to heal, forgive, and find the strength to continue the fight for justice. It is for the White person who fears being labeled a racist and is silenced but yearns to speak up and walk alongside traditionally marginalized groups as they journey towards healing. We wrote this book so readers can deeply understand our shared humanity and the positive outcomes that are possible in our organizations and lives when we stand united and resolve to cultivate environments of belonging for all identity groups.

Following the unfortunate and violent deaths of George Floyd, Breonna Taylor, Ahmaud Arbery, and many others, we found ourselves in many conversations with leaders across the country wondering how to lead their cities and organizations through the troubling times ahead. These leaders were criticized for speaking up, and they were also condemned for being too quiet. It seemed as though they could not do anything right. As we sat with them and talked through their hopes and fears, we were able to discuss the concept of shared humanity, and it seemed to resonate. It provided a frame of reference that was desperately needed and gave them answers on how to take their next steps.

In our conversations, we spoke about the way our brains and minds function and how and why human beings react in specific ways to situations, especially when they feel unsafe, scared, and devalued. We also talked about how people react when they feel stifled in finding and achieving their life's callings. As we talked, it became evident that many White people we spoke with were able to perspective-take in a way they had not been able to before; they reported that their capacity for empathy had increased.

A White organizational leader explained as he reflected on a conversation we had:

> *I still remember that day, June 22, 2020, when our community was torn apart by the turmoil. In that moment, you brought hope and possibility, and we cannot thank you enough for what you did for our community.*

Another White organizational leader explained:

I have found a way to participate, contribute, have a voice, feel connected, and belong.

The people of color we spoke to were able to make sense of their emotions and behaviors and were empowered with language to articulate how and why they felt a particular way. A Black community leader told us:

I want you to know that you have really touched our family at home. I want to let you know that everything I have heard so far has taken me on some type of time travel back through my life, back through my childhood. I experienced everything you have spoken about, and I can remember specific instances throughout my entire life at school—with friends, as a cheerleader, and at work. I didn't know names for a lot of things that were happening and those things I was experiencing, and I didn't understand how to articulate things with my own children...

The overwhelming feedback we have received has fueled our passion for unity and reconciliation and motivates us to continue the work of reminding people of our shared humanity.

The perspective of shared humanity clarifies the need and the power of human connection and the importance of working towards unity and reconciliation while breaking down barriers in an interdependent manner. Shared humanity compels a call to empathy and compassion with action. It forces us to listen well to each other, appreciate one another's perspectives, and learn to support those around us.

Whether you are in business, education, or a religious organization, whether you are part of a social group, or whether you are just an individual looking to be an influencer and change agent in the work of Diversity, Equity, Inclusion, and Belonging (DEIB), this book is for you. We hope you will read this book and pass on the information you learn to your children so they can live in a world marked by love, compassion, and empathy—one where they recognize the power of unity.

Here's a summary of each chapter:

Chapter 1: Understanding Our Shared Humanity

This chapter explains how the feeling of inclusivity and a shared existence is built into us physiologically. We are hardwired for interdependence to survive, beginning with the lowest level of our biology, the atom. We will share the work of psychologist Abraham Maslow and how the hierarchy of human needs motivates us to fulfill our desire to belong. In addition, we explore the negative impacts of not belonging, followed by suggestions on how we can make better decisions individually to allow ourselves to feel included.

Chapter 2: The State of Hyperawareness

In this chapter, we discuss a state of mind called high-context dependency. When individuals enter this state, they pay close attention to others' comments and behaviors. When in a new or uncertain environment, the brain, where high-context dependency comes into play, will automatically ask four questions to assess risk or threat. We explain how the thalamus and neocortex process signals that cause humans to react a certain way. The high-context-dependent state is mediated by repeated exposure to stressful events (direct or indirect), negative biases, stereotype threat, imposter syndrome, uncertainty, and unbelonging.

Chapter 3: Biases

This chapter discusses how bias is a significant barrier to feeling we belong in our organizations. Often we perpetuate bias verbally, other times through actions. Our brains filter those negative words and acts, causing us to react in specific ways. Both our prefrontal cortex and the amygdala play a role in this process. We will explore how and why our brain processes signals and information the way it does and why we respond the way we do. In addition, we will explain elements of biases, including individual bias, affinity bias, diversity of social networks, and the negative aspects of bias. The offered solution is to build rapport, establish personal contact, validate others, become self-aware, engage intentionally, and remain positive.

Chapter 4: Organizational Bias

This chapter explores how biases impact employees, customers, or students in their day-to-day lives. We will illustrate how our minds constantly try to classify and make sense of things because the human brain does not function well with uncertainty or ambiguity. When the brain detects a situation as uncertain, it will render it unsafe and a potential threat, explaining why we react to new situations the way we do. The remedies for organizational bias will involve a detailed step-by-step approach to how organizations should create inclusive environments.

Chapter 5: Binary Thinking

When we exist in a polarized and divisive context, it compromises trust between people within organizations. Unfortunately, in society today, many tend to overlook our common ground, not employ compromise, or are unwilling to dialogue about our differences. Instead, we look to point out the worst in one another while transforming into a "winner takes all" society—pursuing our self-interests and reinforcing our polarized positions. This is binary thinking. The chapter will also explore the "stakeholders" in any organization and how they may react to information received from the organization.

Chapter 6: Subtle Communication

This chapter will investigate the brain's role in microaggressions and revisit the amygdala and prefrontal cortex's role in how we respond to microaggressions. We'll look at how signals travel to these areas of the brain and why we tend toward the emotional response before the rational one. In the end, we will look at how to respond to microaggressions if you are either on the receiving or giving end.

Chapter 7: The Fear of Labels

We describe neuroplasticity in this chapter and how it pertains to the brain's ability to adapt, reorganize, and evolve in response to the experiences we have from childhood onward. We also examine the stages of child and psychological development from birth to young adulthood and how that inevitably impacts how we see things as adults. We discuss how individuals react to stereotype threats such as fending off the label, invigoration, internal attributions, assimilation, and disengagement. In the end, we offer ways to cope with stereotype threats.

Chapter 8: Saving Face

In this chapter, we delve into the psychology of imposter syndrome. We discuss the effects of imposter syndrome: fear of failure, short-lived success, baseline stress and anxiety, negative self-talk, reduced productivity, increased procrastination, difficulty with collaboration and delegation, defensiveness, externalizing, and need for validation. Yet, even though imposter syndrome is applied across cultures, minorities experience it with greater force in tokenism, stereotype threat, the burden of representation, familial pressures, and biculturalism. We also discuss ways to counteract imposter syndrome.

Chapter 9: The Impact of Status

In this chapter, we discuss power and privilege both in historical and present-day contexts. We detail ways organizations can use wealth, laws and policies, social norms, wellbeing, and collective power to become agents of change in the work of unity and reconciliation.

Chapter 10: The Power of Partnership

This chapter explores environments where individuals can feel seen, heard, and safe. An essential first step as an ally is acknowledging that thoughts and behaviors (conscious and unconscious) can be biased depending on our socialization and frames of reference. We discuss the challenges of allyship and how it must be persistent for positive changes to occur.

Chapter 11: How to Have Inclusive, Open, and Honest Dialogue

In this chapter, we discuss how to have difficult conversations about race. We provide some practical tools you can use to succeed in these conversations.

Chapter 12: Conclusion

This chapter reminds us of the importance of our shared humanity in DEI work.

By reading this book, change agents, leaders, and organizations will be able to have more informed dialogues and create inclusive environments by reviewing and modifying existing protocols and practices. Additionally, they will be able to create a more psychologically safe environment for all. People will understand how their brains process information and create biases and perceptions. Once they understand this, they will be able to deconstruct negative biases and train their minds to rely less on automatic generalizations and get to know people as individuals instead of a group from a specific cultural category. There will be more open and meaningful conversations around DEI topics.

PART I
THE NEED FOR
BELONGING

Chapter 1

Understanding of the Need to Belong: Concept of Our Shared Humanity

"The human need to belong and connect with others is even more fundamental than our need for food and shelter."
—Dr. Matthew Lieberman, Author of *Social: Why Our Brains Are Wired to Connect*

Rhonda's Story

The room quieted as one of the leaders came to the microphone to discuss the company's new DEI training. As his speech ended, he turned to an employee. "Rhonda, as a new hire here and a minority, would you like to share with everyone how you've found things so far?"

Cue the spotlight on the new minority hire! Rhonda sat mute, anxious, and worried. If she answered too honestly or was misinterpreted by her coworkers, her work environment would suffer, so her brain performed a risk assessment. "Everything's great," she said with a forced smile.

Rhonda had not been at the organization long before the leaders had committed to training the staff on being culturally competent and inclusive. A good thing, right? Not exactly. What should have been insightful for her colleagues and comforting to Rhonda turned into something much different as the organization's leaders continually singled her out during the training sessions.

"Rhonda, let me ask you this," her manager Mark started as he paused on one slide. "What do Black women think about this concept?"

Again, she found herself answering in a way that would lessen any pushback. However, after a few sessions, she noticed that some other employees became defensive, made negative comments, and expressed antipathetic emotions because they felt blamed, attacked, and victimized by the seminars. Rhonda did not feel empowered to counteract the negative comments made by her colleagues because she was concerned about maintaining an appropriate workplace identity and getting along with others.

During the first few days on the job, Rhonda was optimistic about working in her new environment and excited about the professional challenge. However, the longer Rhonda remained, the more her optimism waned. She worked hard but could never shake her negative feelings about the culture that brewed around her. Her coworkers were friendly on the surface but standoffish, making her feel like she did not belong. So, as the only Black woman in the office, Rhonda focused on blending in instead of sticking out.

Eventually, Rhonda became fatigued because she believed she had to work harder to counteract those cultural stereotypes. Inwardly, she thought she needed to identify closely with the majority culture in her office—morphing into the social norm to keep the peace. Thus, she assimilated to her environment by wearing her hair straight and buying designer clothes. Even the subjects she broached in casual conversation or when called upon in meetings were tempered to fit the acceptable norm. Inevitably, she misplaced her identity and motivation in the process.

"Rhonda, what's up this weekend?" her co-worker Kevin asked as they left the meeting room.

The seemingly innocuous question about her plans for the weekend gave her pause, prompting a generic response. "Oh, nothing. Probably a bike ride and the usual errands. How about you?" Rhonda responded with the brightest smile she could wear.

She wanted to say she was headed to the city for a culturally diverse festival in the area where she grew up. There would be dancing and her favorite foods, and she was counting the hours until she and her friends would attend. Yet, that gnawing voice in her head worried it would sound too ethnic for her coworker, who probably never did anything like that. She feared his judgment and thus felt even more isolated.

Rhonda felt a need to prove her commitment to her job and the higher-ups. So, she worked long hours and tried to make everything perfect. However, Rhonda did not think outside of the box nor contribute her opinions in meetings. In her attempt to conform to the social norms, some people misinterpreted her actions as a lack of competence, ambition, and preparation. Rhonda purposely did not spend time with her colleagues due to feeling out of place; however, her peers perceived this behavior as standoffish and unfriendly. The stress took its toll on Rhonda's mental and physical well-being.

Some might ask, "Why did Rhonda feel the need to conform? Who cares if others like you?" However, the need to belong is not something we make up for ourselves or something only sensitive individuals concern themselves with. This need to belong is hardwired from birth and manifests itself throughout our existence, whether we admit to its existence or not. Therefore, we must understand what drives those needs before determining how we deal with them. Throughout the chapter, we will revisit Rhonda's situation and explain why she reacts and how you should handle a similar scenario.

Interdependence is a Bridge to Belonging

The proverb "No man is an island" is from a sermon given by the seventeenth-century author John Donne. When we think about how the quote relates to Rhonda, it may become evident that her coworkers, leaders, and organizational culture play an important role in whether she feels she belongs. To be clear on a definition, when we talk about belonging, we talk about individuals feeling as though they are supported, valued, accepted, and connected. Charles Cooley's looking-glass concept suggests that how we see ourselves and behave in different environments is a product of how others see us.[1] The previous point underscores that humans are naturally interdependent on one another for their well-being and identity formation. Moreover, as we look through the lenses of history and biology, the importance and effectiveness of interdependence are highlighted.

The next section will demonstrate how historical leaders realized the power of our shared humanity and began to support marginalized groups.

---

[1] Cooley, Charles Horton. (1902). "Human Nature and the Social Order." (New York: Scribner's). *Social Organization.*

Often, these individuals lived in societies where there was immense prejudice and discrimination against Black Indigenous People of Color (BIPOC) individuals. These individuals found a way to challenge their previous assumptions and perceptions of marginalized people, became agents of change in their era, and began to cultivate a culture of interdependence and belonging.

Examples of Interdependence in History

Throughout history, some individuals who did not share the identity of those who were brutalized and discriminated against, fought alongside historically marginalized groups for racial justice, unity, and reconciliation. We think of examples like John Newton, who initially engaged and profited from the slave trade but continued to be convicted by the brutal treatment of enslaved people and saw their humanity. In the latter part of his life, he became a fierce abolitionist. We consider individuals like John Rankin, a conductor and prominent figure in the Underground Railroad. Rankin worked alongside enslaved individuals, such as John Parker, in Ohio to fight against the institution of slavery. At his funeral, people reported a multicultural celebration where people of different races and backgrounds congregated to honor his work and life. We consider the story of Virginia Foster Durr, a Southern aristocrat who worked alongside Civil Rights leaders Rosa Parks and E.D. Nixon during the Montgomery bus boycotts. She took the leaders into her home and partnered with them on a journey to eliminate racism and discrimination.[2, 3]

Even though we could fill this book with many other examples of how interdependence is essential to racial justice, unity, and reconciliation, it is essential to highlight the opportunity and responsibility we have in cultivating a culture of belonging in the present where we all can thrive.

For those looking to be change agents, whether you are a leader within a business, a religious organization, an educational institution, or an individual looking to make a difference, examples from the past teach us that we should:

- **See the humanity in people**. Look at people as individuals, get to know them, and listen well.

---

[2] Drick, Boyd. (2015). *White Allies in the Struggle for Racial Justice*. Orbis Books.

[3] Brown, Cynthia Stokes. (2002). *Refusing Racism: White Allies and the Struggle for Civil Rights*. Teachers College Press.

- **Exercise humility**. Acknowledge that we do not know all there is to know about a person's perspectives or experiences. As a result, we must be open, teachable, and willing to review our perceptions and assumptions and revise our worldviews.

- **Proximate**. The closer you are to people, the more you can learn from them and find out how to support them. The greater the degrees of physical and identity separation, the more we leave room for assumptions.

- **Understand the biases you carry and educate yourself on how to counteract them**. Even though we all have biases, we can identify their sources, become more self-aware, and work to overcome biases that negatively impact people.

- **Understand the broader goal and vision**. Understanding the broader goal and vision of what you are trying to achieve allows you to play the long game and align your actions to achieve the bigger vision.

Interdependence is an essential prerequisite to creating a culture of belonging within our organizations, social circles, and spheres of influence. It starts with each of us acknowledging the opportunity and responsibility we have to each other. When we are able to operate at our human potential, what we can accomplish in unity is more than we can achieve divided.

Examples of Interdependence in Biology

The biological makeup of life provides essential clues about how the human body and life on Earth are fundamentally built to emphasize group identity and interdependence. Biologists talk about the vertical organization of life, which starts from the most minor aspect of humanity, the atom, to the biosphere, where all the ecosystems on Earth exist.

A single atom is the basic building block for all matter, including human life. Atoms combine with other atoms to form molecules, which provide the basic scaffolding for important macromolecules like DNA, proteins, fats, and carbohydrates. These macromolecules work together to sustain the cell, the basic unit of life. In large multicellular organisms like humans, cells combine and work in concert to form tissues, and tissues make up our organs. Organs work together to make up an organ system, like the nervous system, which is

31

essential for making sense of our internal and external worlds. Organ systems work together to sustain an organism, an individual living thing, such as you and me. A group of organisms of the same type that occupies a specific area is called a population. Cooperation within a population is essential to maximize the resources within an environment. Diverse and different populations inhabiting a geographic area are called a community. Different communities together with abiotic or non-living factors like climate, soil, nutrients, availability, and water make up an ecosystem. Different ecosystems and their interconnectedness make up the biosphere, the Earth.

The intricate and essential connections that extend from the atom to the biosphere do not happen spontaneously but are formed intentionally to achieve a bigger purpose—to survive, thrive, and reproduce. The successful functioning and interdependence of the different molecular and living structures rely on characteristics that both leaders and change agents can learn from. While we will not talk about how these characteristics relate to molecules and cells, we will extrapolate the themes and apply them to you and situations where you find interdependence an essential tool for success. The characteristics for interdependence are:

- **Communication.** People need to be clear and consistent with communication to avoid misunderstandings and assumptions.

- **Understanding how your part fits into the whole**. By understanding how your part fits into the whole, you will be able to appreciate the value you bring to the overall goals of an organization or system you work in.

- **Accountability**. Accountability is important in upholding the mission, vision, purpose, and values of individuals and organizations. Accountability is important to ensure that each person is doing their part and contributing to the success as a whole.

- **Resolving differences and incompatibilities**. It is important to work through conflict productively so that people can move forward together.

These characteristics are crucial to a culture of belonging and may give people an increased sense of belonging and value in an environment.

History and biology effectively teach us, human beings, the power of interdependence. The core theme emerging in the paragraphs above is that interdependence is essential to belonging.

Maslow's Hierarchy of Needs and Its Relation to Shared Humanity

As leaders and agents of change, it is important to consider that we are connected because we share fundamental human needs (our shared humanity). In fact, the beauty of our shared humanity is that we are carefully designed to meet some of these needs cooperatively. Yes! Your family members, colleagues, social peers, direct reports, or bosses are giving you the opportunity to meet some of their fundamental needs when they are interacting with you. Let us explain.

American psychologist Abraham Maslow introduced his theory of human motivation. Maslow asserted that humans have a hierarchy of five basic needs that motivate them:

- Physiological needs
- Safety
- Love and belonging
- Esteem
- Self-actualization

The first four needs are at a lower level on the hierarchy. Physiological needs include air, water, food, shelter, clothing, and reproduction. Safety includes personal security, employment, resources, health, and property. Love and belonging refer to friendship, intimacy, family, and a sense of connection. Esteem is about respect, status, recognition, and power. These needs are broken down further into deficiency and growth needs. Physiological needs, safety, love, belonging, and esteem are deficiency needs. Deficiency needs mean that you are aware of them when they are lacking, and when they are not satisfied, individuals are more motivated to fulfill them. When these deficiency needs are lacking, they can result in negative consequences and harm.[4]

In a previous conversation, Pascal explained Maslow's hierarchy of needs to a mentor, and the mentor commented on how many moving parts there

---

[4] Maslow, A. H. (1948). "'Higher' and 'Lower' needs." *The Journal of Psychology, 25*(2): 433-436.

are to it. As Pascal thought about it, he explained the following, which we have found to be tremendously helpful and advantageous:

- Physical needs correspond to physiological needs

- Psychological needs correspond to (psychological) safety, belonging, and esteem

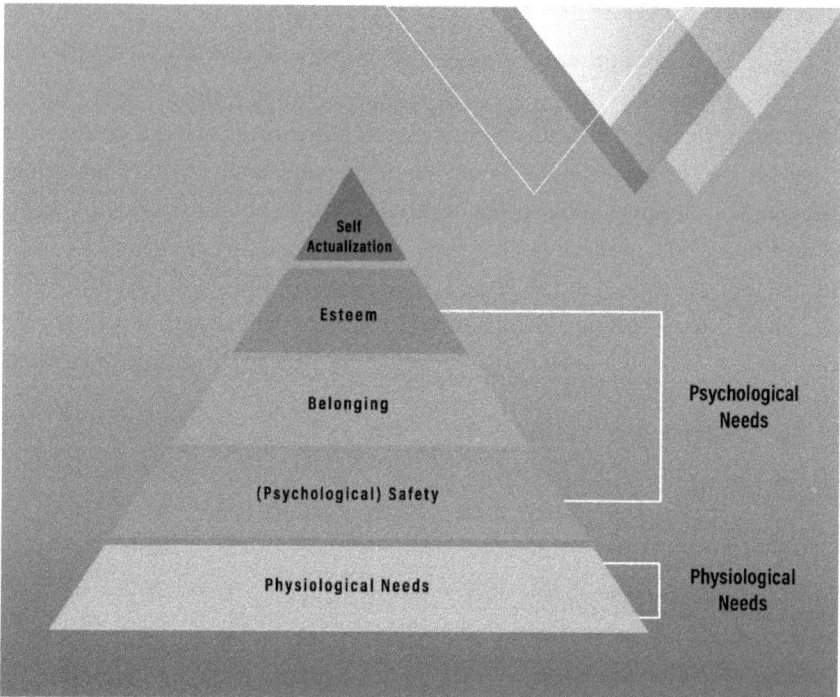

Therefore, if our physical needs, such as food, water, and shelter, are unmet, it could lead to severe physical harm. Likewise, if our psychological needs are unmet, we can experience negative mental well-being, such as depression or anxiety. When our physical and psychological needs are not met, we become aware of them and work to fulfill them, sometimes in unhealthy ways.

Now, let us personalize it for you, the individual reading this book. For a minute, think about the above paragraphs in terms of the need for love and belonging being unmet. What would it mean to you if you did not feel loved or that you belonged? I think each of us can cite a time when we felt that way. Then think of how you reacted. Did you try to meet those desires in other ways? Or have you ever noticed a child acting out or behaving poorly and then heard someone comment, "Oh, they are just trying to get attention"? That is because the child most likely is trying to meet a need, such as love or belonging. Again, as humans, when our needs are unmet, we seek to fill that vacancy.

Self-Actualization

Self-actualization, the fifth and higher-level component of Maslow's hierarchy, is the desire to achieve your life's calling. This is called a growth need, which means it does not stem from a deficiency but rather from wanting to grow, learn, maximize one's creativity, and experience life's pleasures. For example, when we enter an organization many of us consistently look for opportunities to grow, develop, and advance. In fact, for people of color, there are growth needs that directly correlate to the ability of organizations to retain these individuals. They are:

- **Clear communications** about criteria used for promotion

- **Professional development** opportunities to enhance people's skill sets

- **Mentorship and sponsorship** structures

- **The ability to network** with individuals that share similar identities

These factors are connected because they point to the advancement of individuals. Organizations that fulfill the growth needs of individuals in the ways mentioned above were likely to retain their employees of color, particularly Black employees.[5]

There is, however, an important connection between deficiency needs and growth needs that we will discuss in the following paragraph that emphasizes how we should not look at these different needs as isolated but very connected.

Looking Beneath the Surface

Maslow also stated that the four lower-level deficiency needs must be fulfilled before higher-order needs can be satisfied. For example, humans will not focus on internal growth if their basic needs, such as food, safety, love

---

[5] "Being Black in corporate America: An intersectional exploration." (2019). Center for Talent Innovation.

and belonging, or esteem, are unmet.[6] Therefore, if an employee's physical and psychological needs are not met in the organization, they may not grow or reach their full potential. In other words, they will not self-actualize.

Imagine you are in a meeting and feel as though you are not a valued, supported, and connected member of the group. Your leader may ask people to contribute their ideas to an important decision, but the narrative of unbelonging in your mind may cause you to go into self-preservation mode and not say anything. The lack of belonging and trust can easily be misinterpreted as passivity, disengagement, and apathy. While we can ask the employee to push through and find the courage to contribute their thoughts and ideas, the word "culture" stems from the Latin word *colere*, which means "to cultivate," and as a leader you have an opportunity to cultivate an environment of belonging in meetings by doing the following:

- **Providing multiple ways for people to contribute their ideas.** This could be done in written or verbal forms.

- **Allowing for smaller group discussions.** This may allow individuals to contribute their ideas and allow for a representative to share with the larger group.

- **Allowing space for people to process their thoughts.** It is important to understand that different people process things at different rates, so sending prompts ahead of time or providing thinking space during a meeting can be effective.

- **Setting norms for conversations.** It is important to set norms where people understand that it is important for them to share the space with others.

The points above fulfill lower-level deficiency needs and position an individual to have the courage to contribute ideas and have their voice represented in decision making. The more value an individual feels (esteem need), the more they can strive for their maximal and full potential. When someone is achieving their potential, they position themselves for advancement. Imagine a person of color in a culture of belonging. They will

---

[6] McLeod, S. A. (2020, December 29). "Maslow's Hierarchy of Needs." *Simply Psychology.* https://www.simplypsychology.org/maslow.html.

likely achieve their potential and position themselves for promotion, which will likely attract other underrepresented individuals to the organization.

Maslow Before Bloom

In our consulting work with schools across the country, a popular saying has emerged— "Maslow Before Bloom." "Bloom" refers to Blooms Taxonomy, which provides a hierarchy of skills that teachers can implement to enable students to learn and retain information more effectively. When we break down this saying "Maslow Before Bloom", it means that parents and guardians, educators, and community members must work to meet the physiological and psychological needs of students so that they can flourish academically. During the COVID-19 pandemic, many students, especially students of color, experienced food insecurity, and many did not have stable work environments.[7] While teachers and leaders may not be able to alleviate and solve food insecurities, they can make an impact in meeting the psychological needs of students and employees. Some ways to do that include:

- **Taking time to greet students/employees/stakeholders when you encounter them**. A 2018 study showed that for students, positive greetings increased their engagement by twenty percent and reduced disruptive behaviors by nine percent.[8]

- **Enhancing student/employee/stakeholders' resilience toolkits**. This can be done by teaching or encouraging them to use mindfulness, prayer and meditation, breathing, and physical movement to alleviate stress.[9]

---

[7] Haider, Areeba, and Lorena Roque. "New Poverty and Food Insecurity Data Illustrate Persistent Racial Inequities." Center for American Progress, November 5, 2021. https://www.americanprogress.org/article/new-poverty-food-insecurity-data-illustrate-persistent-racial-inequities/.

[8] Cook, Clayton R., Aria Fiat, Madeline Larson, Christopher Daikos, Tal Slemrod, Elizabeth A. Holland, Andrew J. Thayer, and Tyler Renshaw. "Positive greetings at the door: Evaluation of a low-cost, high-yield proactive classroom management strategy." *Journal of Positive Behavior Interventions* 20, no. 3 (2018): 149-159.

[9] Berger, Tom. (2020, September 23). "How to Maslow before Bloom, All Day Long." Edutopia. George Lucas Educational Foundation. https://www.edutopia.org/article/how-maslow-bloom-all-day-long.

- **Gathering regular feedback.** Feedback enables you, as a teacher or leader, to determine how people are experiencing your culture and give them a voice.

The suggestions above work to fulfill the psychological needs of students and employees and position them to maximize their potential in the workplace.

Maslow and Stages of Development

While conducting a training, an organizational leader asked an important question: "When does our need to belong start to emerge?" The need to belong shows up very early in human development, and to drive this home, let us look at Maslow's hierarchy of needs in the context of human development.

- For infants 0 to 1.5 years old, their greatest physiological needs are food, water, sleep, and shelter.

- For toddlers between 1.5 and 3 years old, if their physiological needs are met, their primary need is for safety: freedom from fear, a stable family environment, consistency of relationships, and freedom from danger.

- For preschoolers between 3 and 6 years old, if the above two needs are met, they are looking for love and belonging, which includes being understood and accepted and feeling like part of a group.

- For adolescents and adults, if the above needs are met, their focus is esteem, including desiring status, appreciation, respect, autonomy, and self-worth.[10, 11]

As you may have gathered from the section above, children start to form relationships as they mature and develop. Research by Rekers and colleagues showed that children as young as three years old would rather work

---

[10] McLeod, S. A. (2018, May 3). "Erik Erikson's Stages of Psychosocial Development." *Simply Psychology*. https://www.simplypsychology.org/Erik-Erikson.html.

[11] Over, Harriet. (2016). "The Origins of belonging: social motivation in infants and young children." *Philosophical Transactions of the Royal Society B: Biological Sciences* 371(1686): 20150072.

collaboratively than in isolation to achieve a particular goal.[12] In the preschool years, children form stable friendships and work to maintain these friendships.[13] When children are excluded, it negatively affects their moods, self-esteem, and self-worth.[14]

Negative instances of exclusion stick with us, even into adulthood. Many adults today can pinpoint and describe situations in their childhood where they experience unbelonging or exclusion. The brain has an uncanny ability to hold onto moments of unbelonging and remind us of them when we experience negative situations in our environments. For some, the fear of being excluded follows them like a shadow into joyful experiences and robs them of their ability to flourish in a particular setting.

Because our need to belong is a fundamental human need, the way we try to cope is to deemphasize parts of our identities that are not accepted to assimilate and fit in. Researcher Brené Brown suggests that fitting in or changing important parts of ourselves is at odds with belonging, and so as Maslow suggests, unbelonging will stifle an individual from fulfilling their true potential.[15]

For leaders within organizations that are looking to promote a culture of belonging versus fitting in, here are a few things for you to keep in mind:

- **Ensure that your definition of diversity is broad.** When your definition of diversity encompasses many identity groups, especially those in your organization, stakeholders may feel seen and valued.

- **Use selective vulnerability.** We all have a deep desire to belong, and as a leader you can set the tone by sharing your stories of unbelonging and how organizational culture plays a role in promoting belonging.

---

[12] Rekers, Yvonne, Daniel BM Haun, and Michael Tomasello. (2011). "Children, but not chimpanzees, prefer to collaborate." *Current Biology* 21(20): 1756-1758.

[13] Hartup, Willard W., and Brett Laursen. (1999). "Relationships as developmental contexts: Retrospective themes and contemporary issues." In *Relationships as Developmental Contexts*, 27-50. Psychology Press.

[14] Nesdale, Drew, and Anne Lambert. (2007). "Effects of experimentally manipulated peer rejection on children's negative affect, self-esteem, and maladaptive social behavior." *International Journal of Behavioral Development* 31(2):115-122.

[15] Brown, Brené. (2010). *The Gifts of Imperfection: Let Go of Who You Think You're Supposed To Be and Embrace Who You Are.* Simon and Schuster.

- **Collect data.** Use data, like anonymous surveys, to determine how different groups are feeling within the organization.

- **Have clearly defined anti-harassment and anti-discrimination policies.** This may show an organizational commitment to different identity groups.

Rhonda's Story

Now think back to the story at the beginning of the chapter with the Black female employee, Rhonda. If individuals walk into an organization and feel they do not belong, they will see their differences as obstacles and spend their energy trying to fit in and find their place. For Rhonda, this was evident when she straightened her hair, spoke in a certain way, and tried to maintain the image of the model minority. The tragedy is that Rhonda never reached her full potential in that job because her priority was achieving her lower-level needs like psychological safety, belonging, acceptance, and assimilating to the dominant culture. Ultimately, she left the company when those needs went unmet. Rhonda would have likely exhibited creativity, motivation, and engagement in a better, more inclusive environment where she felt supported and accepted.

The Impact of Negative Emotions

Even though Rhonda was initially optimistic, she felt isolated and excluded in her new environment. When her need for belonging was unmet and the stress of assimilation became too much, Rhonda developed significant anxiety and felt better off leaving her job. Therefore, lacking something we intrinsically need creates a negative emotion with potentially damaging consequences. Let us explain why that point is critical.

Humans have six to ten basic emotions: anger, contempt, disgust, distress, fear, guilt, interest, joy, shame, and surprise. These emotions serve a broader purpose both evolutionarily and in contemporary society. Specifically, emotions are largely unconscious and automatic. We do not choose to feel them, but they allow us to recognize and react to situations in our world.[16]

Evolutionarily, negative emotions cause us to distance ourselves from a stressor or danger as a means of survival. We are wired to find the negative

---

[16] Ekman, P. (1992). "Are There Basic Emotions?"

in different situations to protect ourselves emotionally and physically. Thus, from the beginning of time, humans have always approached a new environment focused on the negatives first as a means of self-protection.[17]

Additionally, when individuals feel like they do not belong, they are hypersensitive to negativity. Therefore, when they enter an environment assessed as hostile, their mental energy is spent performing threat appraisals of that environment because they do not feel safe, accepted, or supported.[18] For example, research shows that individuals who have had a significantly negative experience in an organized program look for and remember negative behavior and facial expressions.[19] They may also interpret well-meaning and constructive feedback as an attack rather than an attempt to help. As a result, an employee may not feel safe in the organization, hampering their performance. Again, we saw this in the story of Rhonda. Remember, these reactions are all biologically based.

Counteracting Negative Emotions

Think about a job where you did not believe you fit in or a school you attended where you felt like a fish out of water. What happened when you made that first good connection with someone? Did each positive interaction from that point move you away from those feelings of uncertainty and worry? They probably did because with each new productive interaction, you felt like you belonged a bit more, one positive emotion after another to counteract the negative.

We cannot overemphasize the power of positive emotions. Researcher Barbara Fredrickson introduced the concept of the "broaden and build" theory. Through a series of laboratory-controlled experiments, Fredrickson explained that positive emotions, such as gratitude, joy, awe, serenity, interest, hope, pride, amusement, inspiration, and love, can increase our intellectual resources and undo fatalistic thinking while building resilience. When we take a moment to channel positive emotions, especially in a stressful situation, it

---

[17] Vaish, A., Grossmann, T., & Woodward, A. (2008). "Not All Emotions Are Created Equal: The Negativity Bias in Social-Emotional Development." *Psychological Bulletin*, 134(3), 383.

[18] Walton, Gregory M., and Geoffrey L. Cohen. (2007). "A question of belonging: race, social fit, and achievement." *Journal of Personality and Social Psychology* 92(1): 82.

[19] Bermudez-Rattoni, Federico. (2007). "Adrenal Stress Hormones and Enhanced Memory for Emotionally Arousing Experiences." In *Neural Plasticity and Memory*, 287-306. CRC Press.

enhances our mental resources, such as problem-solving and our capacity to learn. In addition, positive emotions, such as those listed above, allow us to form solid relationships and be open to new ones. Finally, positive emotions can return individuals to a normal physiological state following a stressful or anxious situation.[20]

Furthermore, Barbara Fredrickson's research reinforces this point. She found that it takes more positive emotions to counteract one negative one. Suppose an individual does not get positive reinforcement or affirmation in a work environment. They will see that environment as unfavorable, feel unsafe, and lack a sense of belonging. The worker-employer marriage may fail if those emotions or responses are not counteracted. Yet if they are counteracted with positive reinforcement, the worker may be better able to focus on higher aspirations.[21]

It is important to note that a healthy environment and high belonging does not preclude tension, stress, disagreements, prejudice, discrimination, and negative emotions. Instead, it indicates that there are systems to identify, guard against, respond to, learn from, and move the organization, social group, or friendship forward. Although those systems may never eradicate all issues, the more an organization can lessen them, the happier and more productive their workers will be. If there had been a system to address Rhonda's issues without fear of recrimination and further isolation, she would have had more self-esteem and an increase in her innovative creativity and productivity.

The Exit Interview

As we reflect on Rhonda's story, we imagine a conversation with Rhonda and her manager that may go something like this:

Manager: "Well, I am sorry to see you go, Rhonda. We barely got to know you. Just out of curiosity, what caused you to start to look for other opportunities?"

Rhonda: "I appreciate you asking. I just didn't feel like I had what I needed to succeed here."

---

[20] Fredrickson, B. L. (2013). "Positive Emotions Broaden and Build." *Advances in Experimental Social Psychology*, 47: 1-53. Academic Press.

[21] Crosby, J. R., Monin, B., & Richardson, D. (2008). "Where Do We Look During Potentially Offensive Behavior?" *Psychological Science*, 19(3): 226-228.

Manager: "Could you help me understand what success would have looked like to you?"

As we think of what Rhonda's response would be, we cannot help but reflect on our own experiences. We feel exhausted when we try to explain ourselves while grappling with the possibility that someone will not understand our experiences as people of color. We also experience the day-to-day stress of having to consider whether we will encounter an instance of bias or microaggressions in the workplace. The exhaustion stems from the constant self-doubt as to whether we are overreacting to situations and whether we should stay silent, put our heads down, and do what was asked of us without complaint. The exhaustion comes from figuring out the parts of ourselves that are accepted and the parts that we must deemphasize to fit in. The exhaustion comes from potentially putting on a brave face after witnessing brutal acts of violence against people that share an identity we hold dear.

We feel for Rhonda as she grapples with the version of herself she needs to put forward to answer the question the manager posed. Maybe Rhonda will code-switch, adjust her behavior to make her manager feel comfortable, and not disclose her feelings and thoughts. She may act like this because of the fear of being labeled "the angry Black woman" and risk having a mediocre recommendation for the next role she pursues. Maybe she is exhausted enough that she feels the need to be true to herself and disclose her inner thoughts and feelings.

Manager: "I recognize that I am putting you on the spot with that question. I will ask that if it is a matter of race, I would like to listen, learn, and understand."

There are times when we speak to people who are White and they explain that their exhaustion stems from the fact that they have to shut off their own humanity and feelings for fear of being criticized and called racist. They feel exhausted from knowing they had good intentions while recognizing the unintentional impact of their actions on an individual. The exhaustion emerges from sitting in a petri dish of shame and silence, not knowing what to say and do but yearning to make a difference.

We recognize that these may be sentiments that the manager may be feeling as well: wanting to make a difference while recognizing the unintentional consequences of trying to do the right thing.

Dr. Pascal & Crystal Losambe

As we grapple with the situation described above, we are reminded of a simple truth—there is symmetry in every event. Despite what situation we find ourselves in, we have agency and responsibility to, as Victor Frankl said, choose our attitudes and our reactions. We do not automatically adopt the posture of a victim but continue to fight for control of our thoughts, mindsets, and behavior despite the actions of others. In exercising our personal power, we are in no way condoning what others do or say but are taking control over the environment we live in.

Anyone's journey toward belonging requires that they know themselves and their triggers. In different situations, you must become aware of your emotions and behaviors, either conscious or unconscious. Yet, even in understanding yourself, feelings of unbelonging might break through.

Also, it would be a mistake to think that battling unbelonging is solely an individual's responsibility. In fact, it takes more than you to recognize your reactions and alter the outcomes to enhance or make a situation more favorable. An organization or social group must also create structures, processes, and practices that can position individuals to feel a sense of value, support, and connection. And by organizations we are referring to businesses, schools, houses of worship, government, and communities.

We hear the leader, teacher, pastor, and friend asking us, "How do we do that when we are wounded or are sitting in shame over and over again?" To that, we say that there is a personal and collective cost to divisiveness and unforgiveness. When we react in certain ways to different situations, our brains create neural pathways that get strengthened when we consistently take a certain posture. Over time, the reaction controlled by that neural pathway may become our default, and it may become difficult to choose a different route. However, it is still possible given the neuroplastic nature of the brain. It may be essential to play the long game and continually remind ourselves of the end goal: belonging, unity, and reconciliation. Are we willing to give up short-term satisfaction for long-term rewards? If we reflect and write down our long-term goals, the mind and brain will become sensitive to our actions and behaviors in relation to the overall goal. We are not saying that you will not experience anger, frustration, disgust, anxiety, and fear, but you are more apt to rebound and reflect on ways to correct your behavior when you find yourself in a similar situation.

If racial justice, unity, reconciliation, and belonging are the goals, then we all belong in this work. If we are all affected to varying degrees by the realities and legacy of racism, then we all belong in this conversation. In the midst of divisiveness and unprecedented polarization, a third narrative must emerge—a narrative where we elevate our shared humanity and recognize the human potential we have. Armed with humility, compassion, empathy, dignity, and respect for each other, we can start to leave our society and world better than we found it.

Final Reflections on Rhonda

As we turn back to Rhonda's situation, it may be essential to determine what happened with her physiological and social levels. Like any of us, Rhonda had a fundamental need to belong. She was likely sensitive to signs of unbelonging, especially in her new environment. The various situations she found herself in resulted in stress, which caused her to hyperfocus on the things that contributed to her not belonging. As Rhonda persisted in that environment, her brain had likely created a generalized organizational narrative that said, "You don't belong here."

Once the brain forms a narrative, especially a deficit one, it becomes hypersensitive to behaviors or actions confirming that narrative. In addition to trying to settle into her role, Rhonda may have contended with whether she belongs to her new organization or not. There is a cognitive cost to unbelonging because Rhonda's mental capacity has likely undergone threat appraisals in her environment while trying to focus on her assigned tasks.

Rhonda likely wanted to feel accepted and be honest about how she felt, which would have activated the reward pathways of her brain and allowed for positive engagement, motivation, and contribution. The time and energy Rhonda spent trying to fit in and belong depleted her productivity and ultimately impacted the organization's goals.

So, there is a lesson to learn here for the organization. There are a few ways the organization could have increased Rhonda's sense of belonging:

- **Status (Am I valued?)** Understand what is important to Rhonda, give positive feedback, and make her feel like an integral part of the team.

- **Certainty (Do I know the expectations?)** Provide clear expectations for Rhonda's behavior and open, transparent communication.

- **Autonomy (Do I have a voice?)** Have structures available for Rhonda to share her voice and give perspective on what may or may not be working well.

- **Relatedness (Do I have meaningful connections?)** Create onboarding processes that enable Rhonda to meet her colleagues and learn the history and values of the organization.

- **Fairness (Am I respected and treated with dignity?)** Assist in Rhonda's feeling that she is nurtured, developed, and invested in as much as other people in the organization. There are clear pathways for promotion and transparent compensation.[22]

Summary

Here are ten takeaways about how you can help build a culture of belonging.

1) Build diverse teams, defining diversity as broadly as possible.

2) Develop a learner's mindset, always being open and humble about others' experiences, especially knowing you will discover your own biases.

3) Keep the broader vision and your part in it in mind.

4) Communicate to understand and work through differences.

5) Put concrete organizational supports in place for people of color, including:

- Clear communications about promotion criteria

- Professional development opportunities

- Mentorship programs

- Networking systems for people who share identities.

6) Have clearly defined anti-harassment policies.

---

[22] Rock, D. (2008). "SCARF: A Brain-based Model for Collaborating With and Influencing Others." *NeuroLeadership Journal*, 1(1): 44-52.

7) Provide multiple ways for people to contribute.

8) Implement feedback into operations and model accountability.

9) Lead with selective vulnerability.

10) Establish personal connections with stakeholders, including greeting them by name.

PART II
BARRIERS TO
BELONGING

Chapter 2

The State of Hyperawareness

"Ubuntu really says if you want to be nice to yourself, start in a way by being nice to the other."
—Desmond Tutu

I magine the trepidation of one's first day at a new job, social group, or school. We have all experienced this fear on some level. Now read the following scenario:

A recent graduate from a local, well-respected university is hired as an accountant for a large corporation. The graduate is more than qualified for the position, but she knows no one in the office where she will work. In fact, she immediately learns that everyone is much older than she and has been with the company for years. The new employee finds herself experiencing bouts of anxiety because of the uncertainty she feels going into her new role. She is concerned about whether people will consider her incompetent because of her age. What if she says the wrong thing or her learning curve is too steep? What if she asks too many questions and takes up people's time?

On her first day of work, her manager was kind enough to take her around to each person's desk and introduce her one-on-one. Each person she met was friendly and welcoming but reserved.

When lunchtime came, the other employees got up to go to lunch together and did not ask her to join. In fact, they walked past her desk and did not acknowledge her at all.

Since almost everyone has been the "new person" at some point, we can all identify with a certain amount of apprehension or nervousness you may experience. Plus, each of these situations could provide some increased anxiety. However, as readers, we cannot necessarily understand how each person may process these new situations—individuals bring different emotions to any case. So, where one may be upset, another may not even be affected.

When humans enter a new environment, there is a degree to which they pay attention to another's comments and behaviors. That information is then placed into a larger "context," which gets filtered through one's experiences, values, beliefs, mindsets, norms, and world views before meaning is applied. In other words, context acts as a filter.

Individuals who use many cues in their environment and filter them through their experiences, beliefs, values, and perspectives are in a high-context-dependent state. For example, when individuals enter into the high-context-dependent state, they diligently watch for non-verbal and verbal cues to give meaning to words that are being said and actions that are being performed. The cues can include signals such as body language, how the person is greeted, their physical environment, conversing in a particular space, and what others are saying to them.[23, 24, 25, 26] In fact, research studies show that when we enter an environment where there is high uncertainty, our pupils dilate, let more light in, and make us more sensitive to our environments.[27]

---

[23] Nguyễn, P. M. (2017). *Intercultural Communication: An Interdisciplinary Approach: When Neurons, Genes, and Evolution Joined the Discourse.* Amsterdam University Press.

[24] Leary, M. R. (2010). "Affiliation, Acceptance, and Belonging." *Handbook of Social Psychology,* 2: 864-897.

[25] Leary, M. R., Tambor, E. S., Terdal, S. K., & Downs, D. L. (1995). "Self-esteem as an Interpersonal Monitor: The Sociometer Hypothesis." *Journal of Personality and Social Psychology,* 68(3): 518.

[26] Allen, K. A. (2020). *The Psychology of Belonging.* Routledge.

[27] Lavín, Claudio, René San Martín, and Eduardo Rosales Jubal. (2014). "Pupil dilation signals uncertainty and surprise in a learning gambling task." *Frontiers in Behavioral Neuroscience,* 7: 218.

To illustrate this further, let us assume you are the employee in the scenario at the beginning of the chapter. If you are in the high-context-dependent state, you will pick up on your colleagues' lack of eye contact as they walk by, how warm their responses were when you met them, or how they said hello quickly but then went back to their computer screen. As a result, a high-context-dependent state would leave you feeling unwelcome, and you may potentially see this as a problem. Your anxiety may be higher than when you started the day.

The Brain and High-Context Dependency

The brain is a predictive and anticipatory organ. When you encounter a situation, your brain predicts what will happen next, creating a response to a particular behavior.[28] The brain does this by filtering information received through a person's past experiences, through their values and beliefs about the world. When individuals are in an unfamiliar situation, they naturally take cues from others' actions or words, filter them through what they know, feel, or have experienced, and apply meaning to them, predicting what might happen next—all these steps occur in the brain.

In any interaction, the brain asks four questions:

1. What are the intentions of this person?

2. Can I trust this person?

3. Is this a safe place?

4. Does this behavior remind me of a situation I have seen or been in before?[29, 30]

Your brain then answers these questions within milliseconds of an interaction. Moreover, these questions serve two main goals: to keep you safe and to maximize rewards.

The thalamus, the coordinating center, receives information from the environment through all senses in the brain, except for smell. The thalamus is connected to the brain's emotional center, mainly through the amygdala.

---

[28] Barrett, L. F. (2020). *Seven and a Half Lessons About the Brain.*

[29] Nguyễn, P. M. (2017). *Intercultural Communication: An Interdisciplinary Approach: When Neurons, Genes, and Evolution Joined the Discourse.* Amsterdam University Press.

[30] Cuddy, A. J., Fiske, S. T., & Glick, P. (2008). "Warmth and Competence as Universal Dimensions of Social Perception: The Stereotype Content Model and the BIAS Map." *Advances in Experimental Social Psychology*, 40: 61-149.

The amygdala is like a danger or fear detector and the site of fear conditioning where we learn what is dangerous and what we should fear. The signal received through the senses also goes from the thalamus to the neocortex to be processed. The neocortex is where your higher-order thinking resides: your ability to reason, control impulses, break down a situation, and think through long-term consequences.

The distance from the thalamus to the amygdala is shorter than from the thalamus to the neocortex. So, the amygdala, our emotional center, receives the information first. Therefore, we "feel" first and then make sense of that feeling or anticipate how we should react to a situation. The amygdala is connected to the hippocampus, where short-term memories reside. The amygdala checks with the hippocampus to determine whether past experiences and emotions pertain to a particular situation.

If there are negative experiences and perceptions associated with a particular person, identity group, physical space, and so on, your body will likely mount the fight, flight, or freeze response when you have a second experience. This can send you into a state of high-context dependency where you are cautious and hypervigilant in your environment and scan it for threats. So, in a situation where you are asked to do a task that requires you to work with others you have perceived as unwelcoming or psychologically unsafe, you may approach them with hesitation and be on guard.[31, 32, 33, 34]

Additionally, the brain will distinguish individuals as "us" and "them" in high-context dependency situations, creating divisiveness. This process creates barriers to trust, safety (especially psychological safety), and collaboration. Research shows that when we label individuals as "them," we may overgeneralize and infuse a potentially negative bias. We will then more frequently notice negatives about those individuals and may miss out on the value or potential they bring to our lives and our organizations.[35]

We have talked about the first three questions people ask themselves in any interaction: what are the intentions of this person, can I trust this person,

[31] Carey, J. (1990). *Brain Facts: A Primer on the Brain and Nervous System.*

[32] Wolfe, P. (2010). *Brain Matters: Translating Research Into Classroom Practice.* ASCD.

[33] Hammond, Z. (2014). *Culturally Responsive Teaching and the Brain: Promoting Authentic Engagement and Rigor among Culturally and Linguistically Diverse Students.* Corwin Press.

[34] Bard, A., & Bard, M. G. (2002). *The Complete Idiot's Guide to Understanding the Brain.* Penguin.

[35] Casey, M. & Robinson, S. (2017). *Neuroscience of Inclusion: New Skills for New Times.* Outskirts Press.

and is this a safe place? Let us talk more about "Does their behavior remind me of something I have seen before?" and how it applies to high-context dependency.

When we experience things, especially when there is high emotion, we tend to remember our physical surroundings (external context) and physical, psychological, and emotional states (internal context) during the experience, particularly when we have heightened emotions.[36] For example, can you remember where you were and how you felt on September 11th, 2001 or when you heard about or witnessed the murder of George Floyd? You may find you can clearly remember your physical surroundings and physical and emotional states because of the tragedy and subsequent heightened emotions surrounding it.

Understand that when individuals are in a situation that reminds them of a past event they witnessed or directly experienced, they retrieve and recall the external and internal contexts of that previous experience. Research indicates that individuals who display high-context dependency are proficient at synthesizing memories. In addition, they are more adept at grouping memories and forming general impressions than someone not in a high-context-dependent state.[37, 38, 39, 40, 41]

As we mentioned earlier, the brain is a predictive organ, so it anticipates what will happen next and causes you to act to protect yourself and maximize rewards. In fact, the brain sorts out millions of pieces of information at any given moment. The RAS (reticular activation system) part of the brain stem filters out non-crucial information and forces individuals to pay attention to specific bits of information that affirm their values, beliefs, and identity.

For example, suppose you are in a crowded room and people are talking. Suddenly, someone says your name, and you turn and look. Why did you

---

[36] Tulving, E. (1974). "Cue-dependent forgetting." *American Scientist*, 62: 74-82.

[37] Hall, E. T., & Hall, T. (1959). *The Silent Language* (Vol. 948). Anchor Books.

[38] Hall, E. T., & Edward, T. (1969). *The Hidden Dimension.* Anchor Books New York: 20, 71.

[39] Hall, E. T. (1989). *Beyond Culture.* Anchor Books.

[40] Hall, E. T., & Hall, E. T. (1989). *The Dance of Life: The Other Dimension of Time.* Anchor Books.

[41] Lakhani, D. (2008). *Subliminal Persuasion: Influence and Marketing Secrets They Don't Want You To Know.* John Wiley & Sons.

react? Because your name is significant to your identity, and your RAS is sensitive to that.[42]

Context-Dependency and BIPOC Experience

Uncertainty has historically been prevalent for many people of color. Individuals were not sure whether the family unit would remain together. There was uncertainty around whether people would assign negative stereotypes to individuals and look at them from a deficit point of view. As a result, many people of color may avoid uncertainty, become observant in different environments, and feel comforted when certainty in an environment is present.

Research indicates that many people of color find it essential to have autonomy in defining, naming, and speaking for themselves. This autonomy gives them a sense of control over their internal states and allows them to better connect with those in their environments.[43] However, in U.S. history, many people of color did not have a voice or contribute to the landscape of the nation they were part of. They were segregated from the majority culture and unable to vote. They were also considered property and could not control their destiny because they depended on others to determine their future.[44, 45]

So, what types of things, mindsets, and values are humans most sensitive to, especially people of color?

- **Status:** Status is defined as one's sense of importance and value. This notion is driven by the question, "Do I matter to the organization?" The idea of status is crucial to many people of color in the U.S. due to the historical realities of being treated without dignity, respect, and dehumanization. As a result, many American people of color are sensitive to behaviors and actions that negatively affect their dignity and self-worth.

---

[42] Garcia-Rill, E. (2015). *Waking and the Reticular Activating System in Health and Disease.* Academic Press.

[43] Van der Kolk, Bessel. (2014). *The Body Keeps the Score: Mind, Brain and Body in the Transformation of Trauma.* Penguin UK.

[44] Duncanson, K. M., Major-Donaldson, B., & Weekes, E. (2016). "Hofstede Theory and Subcultures." In *Allied Academies International Conference. Academy of Organizational Culture, Communications and Conflict Proceedings,* 21(1): 11. Jordan Whitney Enterprises, Inc.

[45] Chang, E. C., Downey, C. A., Hirsch, J. K., & Lin, N. J. (2016). *Positive Psychology in Racial and Ethnic Groups: Theory, Research, and Practice.* American Psychological Association.

- **Certainty:** Certainty is one's sense of clarity and predictability in an environment. Individuals in a high-context-dependent state find uncertainty threatening and problematic. The brain loves patterns and predictability because it provides comfort, and it does not have to spend as much energy doing threat appraisals.

- **Autonomy:** This is the ability of an individual to contribute to an organization and influence the organizational culture. These individuals may ask: "Can I bring creativity to a project, have the flexibility to accommodate unexpected circumstances, and be more myself in the workplace?"

- **Relatedness:** An individual's sense of connection is their ability to relate. We are social beings made to connect; it is a fundamental need for our humanity. Feelings of unbelonging and not being our authentic selves because of stereotype threat and imposter syndrome can influence an individual's sense of connection. For people of color, particularly African Americans, communal encouragement, spirituality, and interconnectedness are essential to their identities.[46]

- **Fairness:** Differential treatment and unequal access to resources and opportunities are considered unfair. When we consider the historical realities of unequal opportunities, whether it was housing, wealth, education, services, and so on, the treatment of people of color led to inequitable outcomes.[47]

Therefore, leaders within an organization need to be consistent, communicative, and dependable with high-context-dependent employees, members, students, and stakeholders. This creates a safe environment for people and enhances their ability to participate at a high-performance level. Remember from the prior chapter that if individuals are focused on a basic need such as feeling safe, their ability to perform or contribute will be diminished.

---

[46] Chang, Edward Chin-Ho, Christina A. Downey, Jameson K. Hirsch, and Natalie J. Lin, eds. (2016). *Positive Psychology in Racial and Ethnic Groups: Theory, Research, and Practice.* Washington, DC: American Psychological Association.

[47] Rock, D. (2008). "SCARF: A Brain-based Model for Collaborating with and Influencing Others." *NeuroLeadership Journal*, 1(1): 44-52.

For example, a manager asks each team member to solve a company problem and explain their answers in a short essay. After reviewing the team's work, the manager picks one paper off the stack and raves about the ideas. He then looks at the name on the paper and calls the name out. An African American employee stands up. When the manager sees who it is, he looks at the man with shock and a measure of doubt and asks, "You came up with this?" Now, the manager's intention may not have been to offend. However, the statement's impact could potentially cause the employee to feel devalued and insulted.

Therefore, when we interact with others, especially those from diverse backgrounds, we must consider the intentions behind our behaviors and the impact they could have. We are responsible for using words and actions that project inclusivity and avoiding appearing judgmental. Again, the manager may not have meant it that way, but his reaction and words told the employee something different.

Intentions are invisible. Actions, words, and behaviors are more visible and therefore tangible. Our verbal communications and the way we present ourselves can have a lasting effect on the person we direct them toward. Individuals in leadership roles must be mindful of these projections to create an environment that promotes trust and psychological safety.

Cues

It is essential to understand that when you interact with others, the recipients of what you say and do are typically attempting to answer four main questions:

- What are their intentions in this interaction?
- Can I trust them?
- Is this a safe place?
- Does their behavior remind me of a situation I have been in before?

These questions are answered mainly using non-verbal cues, although verbal cues serve an essential purpose. Let us go back to the previously described scene. The comment "You came up with this?" will be placed within a larger context, meaning it will be filtered through the employee's experiences, values, beliefs, mindset, norms, and worldviews before meaning is given to the comment. For example, the manager speaking the words and the individual listening will process their interpretations based on what they have individually experienced.

Again, individuals who use many cues, verbal and non-verbal, are considered to be in a high-context-dependent state and will look for eye contact, smiles, body language, tone of voice, the speaker's status, physical environment, and potential differentiation in treatment. In addition, these individuals will also analyze how they are greeted, who is talking to each other, which racial and ethnic groups are present, what or who is represented on the walls and brochures, and other cues in the environment. They may ask whether the staff is paying more attention to others than they are to them.

Pascal was interacting with a friend and explaining the concept of context dependency. She asked him an excellent question that you may be asking right now. She asked, "Is high-context dependency situationally dependent and also dependent on our emotional states?" Good question. Pascal's answer was, "Yes, it is." For example, when we are in a place, space, or interaction where we do not feel valued, affirmed, and supported, we become more highly attuned and hyper-aware of our surroundings. Additionally, our negative emotions cause us to easily recall situations and emotions where we have experienced unbelonging. In those situations, the "threat" switch is turned on in our minds and brain, and we use more environmental cues to assess whether the situation will cause us harm. Different individuals may be more context-dependent than others, especially due to our socialization, backgrounds, and the historical contexts of identities that are important to us.

Pascal was training a group of adults on the subject of high-context dependency. One of the participants made a meaningful connection. The female participant talked about a time she went hiking with a male friend. Before they started the hike, they had a destination in mind. However, as the sky darkened, she began to feel increasingly uncomfortable. She indicated to her male friend that she preferred to turn around and head back to their vehicles because it was starting to get dark. Her male friend responded with his desire to complete the hike. As the female participant reflected, she indicated that she had always been taught to be cautious when it was dark because of any potential dangers, especially for a young woman. Her context caused her to have a different experience and become more hypersensitive to the environment. In contrast, the male felt more secure in that environment and was less sensitive to being out after dark.

Research shows that if individuals or groups encounter discrimination, exclusion, and marginalization repeatedly and see others who share their identities experience the same, their brains make them more hyperaware of their environments. Consequently, ethnic and racial minority populations tend to enter the high-context dependent state more often than individuals that are part of the dominant culture or individuals that have not experienced prejudice and discrimination based on their salient identities, such as race or ethnicity.[48, 49]

The many BIPOC people we have encountered have stated their racial identity is often at the forefront of their minds. When they enter a high context-dependent state, they wonder whether their racial identity will be used as a basis for discrimination and marginalization due to historical and present-day examples of racial injustice. In fact, one study showed that 70% of African Americans are concerned that they or their loved ones will be discriminated against in the workplace compared to 23% of White individuals.[50]

For example, many ethnic and racial minorities, especially African Americans, are taught to be mindful of their presence, dress, and conduct in particular spaces. Many young Black children are given "the talk," where they are instructed to be wary while driving in different neighborhoods or in the presence of law enforcement. Where did this come from, one may ask? Simply stated, it comes from years of either being the target of profiling or hearing and knowing of people in their racial identity who are being targeted.[51]

Neuroplasticity

Some years back, Pascal was debriefing with some students following the movie, *The Hate U Give*. As they talked about various themes in the film, a young Black male spoke about his experience with "the talk." He was about

[48] Soto, J. A., Dawson-Andoh, N. A., & BeLue, R. (2011). "The Relationship Between Perceived Discrimination and Generalized Anxiety Disorder among African Americans, Afro Caribbeans, and non-Hispanic Whites." *Journal of Anxiety Disorders*, 25(2): 258-265.

[49] Jana, T., & Mejias, A. D. (2018). *Erasing Institutional Bias: How to Create Systemic Change for Organizational Inclusion*. Berrett-Koehler Publishers.

[50] Jana, T., & Mejias, A. D. (2018). *Erasing Institutional Bias: How to Create Systemic Change for Organizational Inclusion*. Berrett-Koehler Publishers.

[51] Chang, E. C., Downey, C. A., Hirsch, J. K., & Lin, N. J. (2016). *Positive Psychology in Racial and Ethnic Groups: Theory, Research, and Practice*. American Psychological Association.

to get his license and talked with his parents about how excited he was to drive fast, pick up his friends, and play loud music in the car. While it was clear that he was light-hearted and joking, his parents did not interpret it as such. Instead, they emphasized his need to be mindful of calling too much attention to himself as a Black man. Even though this young man's family was wealthy, his class or financial status could not protect him against the assumptions people could make about him based on the color of his skin. As we sat in the room, one could see the White students' disbelief because they had never heard of "the talk," let alone connect with the need to be hypervigilant in one's environment. However, the White students listened and showed empathy and compassion as they sat and heard the different stories and connections to the movie.

The beauty in this exchange came in the sharing of the experience. Of course, not everyone in the class had the same reference point. Still, everyone left that class with an elevated insight into one another's reference points and why they may process situations differently. Even though the magnitude of these experiences differs with age, children tend to have more neuroplasticity than adults. Neuroplasticity is the brain's ability to form new connections, especially in response to learning. As a child, you learn to do many different things, and connections are made in the brain with each new experience. This allows a child to change their perspective more quickly on various issues and social constructs.[52, 53]

The scenario about the students in the paragraph above is a great example. The White students were more quickly able to show empathy and understanding for their minority counterparts after hearing about their experiences, which will help them change their perspective on social norms. This is also true when protecting oneself for fear of discrimination or marginalization, but it is more difficult to change these perspectives as an adult. As we enter adulthood, we have already confirmed our values and beliefs. Those connections are reinforced with every external exposure, causing us to see the world through a particular lens. Therefore, if you have been discriminated against or have seen loved ones or those within your

---

[52] Johnston, M. V. (2004). "Clinical Disorders of Brain Plasticity." *Brain & Development*, 26(2): 73-80. https://libdatabase.newpaltz.edu/login?url=https://search-proquest-com.libdatabase.newpaltz.edu/docview/71743981?accountid=12761.

[53] Simona Fiori, Andrea Guzzetta. (2015). "Plasticity Following Early-life Brain Injury: Insights from Quantitative MRI." *Seminars in Perinatology*, 39(2): 141-146, ISSN 0146-0005, https://doi.org/10.1053/j.semperi.2015.01.007.

ethnic group be discriminated against, your brain may be conditioned not to trust.

However, it is essential to note that when we talk about ethnic and racial groups entering the high-context-dependent state frequently, it does not mean that all individuals of a particular race or ethnic group will behave a certain way. It would be a mistake to assume that an individual from a traditionally marginalized group is highly context-dependent because that might be far from the truth. This work is not prescriptive but descriptive, meaning that learning these concepts will enable you to explain behavior rather than predict behavior. The more you understand someone else's behavior, the more empathetic and inclusive you can be toward that individual.

Triggers for High-Context Dependency

Whether or not an individual enters a high-context dependent state depends on how they view themselves, their relationships with individuals in that space, or past experiences they may have endured. It also pertains to the people they love or those who share a particular identity and what they have observed or even experienced themselves. Yet, there are also a few common triggers:

- **Repeated exposure to stressful events (direct or indirect):** Repeated exposure to events, circumstances, or behaviors that affect individuals and their well-being, especially if left unaddressed or unresolved. These situations can lead to chronic stress and poor health. Additionally, an individual becomes hypervigilant and hypersensitive, especially in environments with uncertainty and unpredictability.

- **Negative Bias:** When individuals perceive that someone has prejudice, which is negative assumptions and attitudes towards them based on their membership to a particular group in an unfair or unjustified way, it may cause an individual to go into self-preservation mode and enter a high-context-dependent state.

- **Stereotype threat and imposter syndrome:** When individuals are under stereotype threat or imposter syndrome, they may be more apt to enter a high-context dependent state. We will discuss this more in later chapters.

- **Uncertainty:** When there is uncertainty in the environment, individuals can possess high-context dependency, disengage, and go into self-preservation mode.

- **Unbelonging:** When people do not feel connected and accepted in a particular space or by certain people, they can enter a high-context-dependent state.

The Good and the Bad

The high-context-dependent state is not necessarily a bad thing. Again, your brain is trying to keep you safe by undergoing environmental threat appraisals. However, renowned neuroscientist Lisa Feldman Barrett explains that your body has what she terms a "body budget." The activities that make deposits to or increase your "body budget" are hydration, sleep, exercise, connections, gratitude, joy, and so on. However, the high-context-dependent state takes away from or reduces your "body budget." Therefore, as your brain is performing threat appraisals and gathering and storing new information in an unfamiliar situation, it uses resources that could be used for creativity, innovative thinking, completion of a task, or helping others, potentially harming your productivity.[54]

Is There an Answer for Organizations?

The simple answer is yes. Organizations can help high-context-dependent individuals through:
- The conscious building of trust
- Creating psychologically safe environments
- Intentionality

Employees, for example, who find themselves in environments with the above three attributes typically show an increase in productivity, have more energy at work, collaborate better with their colleagues, and have lower attrition rates than individuals in organizations without them.

---

[54] Barret. L (2020). "Your Brain Is Not for Thinking: In Stressful Times, This Surprising Lesson from Neuroscience May Help Lessen Your Anxieties." *The New York Times.* https://www.nytimes.com/2020/11/23/opinion/brain-neuroscience-stress.html.

Trust

Trust is defined as a "psychological state where we accept and display vulnerability based on positive expectations of the intention or behavior of another person."[55]

For individuals that enter the high-context-dependent state often, particularly ethnic and racial minorities, their ability to trust may be slow. They are more hesitant to share information and are cautious about their interactions. Again, this may be due to their experiences and the repetition in which they occur. Therefore, when one's trust is violated, it lowers their ability to be vulnerable and believe that the violator has good intentions or knows how to honor their vulnerability. As a result, these individuals will remain closed off as a means of self-protection.[56, 57]

If trust is what you need to reduce the effects of high-context dependency, how do you create and sustain it? There are four main ways to develop trust within an organization:

- Recognition
- Teamwork
- Autonomy
- Common ground

First, when providing recognition, you need to understand that timing is everything, especially when acknowledging effort and excellence. For example, from employer to employee, recognize a job well done immediately if your employee meets a goal and you see the energy they put forth to complete a task. Then applaud the outcome, process, and character traits it took to achieve a specific purpose. Doing this small task will elevate the person's status and show that you recognize and appreciate their hard work.

When someone feels good about something they may have accomplished, the timeliness of the response to the situation is vital to amplify their

---

[55] Rousseau, D. M., Sitkin, S. B., Burt. R. S. and Camerer, C. (1998). "Not So Different After All: A Cross-discipline View of Trust." *Academy of Management Review*, 23: 393–404.

[56] Kim, P. H., Dirks, K. T., & Cooper, C. D. (2009). "The Repair of Trust: A Dynamic Bilateral Perspective and Multilevel Conceptualization." *Academy of Management Review*, 34(3): 401-422.

[57] Chang, E. C., Downey, C. A., Hirsch, J. K., & Lin, N. J. (2016). "Positive Psychology in Racial and Ethnic Groups: Theory, Research, and Practice." American Psychological Association.

emotions. Moreover, it is essential for the person who needs to develop trust to associate specific behaviors with success, and this will be suggested by the employer to the employee. The employee is looking to answer the following: Are you a joy multiplier, joy thief, conversation killer, or conversation hacker? If someone cannot trust you with good news, they may not feel confident in trusting you with negative information.[58, 59, 60]

Second, when individuals are assigned a challenging task and are asked to work on the job as part of a team, neurochemicals are released: oxytocin and adrenocorticotropin regulate the release of cortisol. The release of cortisol can decrease social connection, focus, and working memory. When a challenging task has an end goal and adequate support is given, trust can grow, and an individual can enter what is known as a state of flow, where they lose themselves in a task and are fully engaged.[61, 62, 63]

Let us go back to the example of the young person just starting their job amongst others who have been at the organization for a long time. Suppose the task given to them by the boss needs help from another employee. If that other employee helps the new worker complete the tasks and shares vital information to support their learning and growth, trust can develop between the two individuals. The new person will relax and feel secure working with that other person. The act of relaxing will allow the brain to focus on the task versus the anxiety or stress. Specifically, individuals tend to gain a sense of trust for others who possess the skills and knowledge of a task and demonstrate a willingness to support and help them.

---

[58] Seligman, Martin EP. (2012). *Flourish: A Visionary New Understanding of Happiness and Well-Being.* Simon and Schuster.

[59] Gable, Shelly L., Gian C. Gonzaga, and Amy Strachman. (2006). "Will you be there for me when things go right? Supportive responses to positive event disclosures." *Journal of Personality and Social Psychology* 91, no. 5: 904.

[60] Woods, Sarah, Nathaniel Lambert, Preston Brown, Frank Fincham, and Ross May. (2015). "'I'm so excited for you!' How an enthusiastic responding intervention enhances close relationships." *Journal of Social and Personal Relationships* 32(1): 24-40.

[61] Zak, P. J. (2017). "The Neuroscience of Trust." *Harvard Business Review*, 95(1): 84-90.

[62] Kosfeld, M., Heinrichs, M., Zak, P. J., Fischbacher, U., & Fehr, E. (2005). "Oxytocin Increases Trust in Humans." *Nature*, 435(7042): 673-676.

[63] Stauble, M. R., Thompson, L. A., & Morgan, G. (2013). "Increases in Cortisol are Positively Associated with Gains in Encoding and Maintenance Working Memory Performance in Young Men." *Stress (Amsterdam, Netherlands)*, 16(4): 402–410. https://doi.org/10.3109/10253890.2013.780236.

Third, many employees desire autonomy, purpose, and a sense of meaning in their work. A 2014 survey by Citigroup and LinkedIn found that almost half of employees would forgo a 20% raise if given more autonomy or control over their work.[64] When teams get freedom as to how they get tasks done, there is an increase in productivity. Independence increases an individual's sense of ownership over a project and shows an increase in intrinsic motivation. Yet, it is essential to note that not all employees like autonomy, so it is necessary to gauge where an individual is on the autonomy scale and spectrum. It is crucial to set the stage and discuss the different approaches with your team and measure the strengths and weaknesses of the approaches.

Finally, trust is built when common ground is discovered. People connect with one another over shared interests and passions. Therefore, it is crucial to get to know people in the workplace, institution, or organization. An example is if you are new on the job and someone comes over to introduce themselves, ask questions, and make a connection. You may discover you both went to the same school or live near one another. Typically, there is an immediate positive feeling at that point, and trust can grow from there.

Psychologically Safe Environments

In 1990, a professor of Psychology at Boston University, Dr. William Kahn, published an article in the *Academy of Management Journal* where he coined the term "psychological safety." Kahn's ground-breaking paper challenged the accepted and prevailing notion that a top-down and hierarchical approach was the way to motivate and engage employees. He asserted that productivity and creativity had more to do with how an employee felt. In other words, did they see purpose in their work, did they feel psychologically safe and psychologically available, or were people cognitively and emotionally withdrawn?[65]

The basis of psychological safety is the individual's freedom to interact without the fear of embarrassment, discrimination, marginalization,

---

[64] Zak, P. J. (2017). "The Neuroscience of Trust." *Harvard Business Review*, 95(1): 84-90.

[65] Kahn, W. A. (1990). "Psychological Conditions of Personal Engagement and Disengagement at Work." *Academy of Management Journal*, 33(4): 692-724.

punishment, or repercussions.[66] Psychological safety works to remove these barriers. The more an individual's thinking, creativity, innovative ideas, and identity can surface to the organization's benefit, the more likely there will be better outcomes for an organization. Conversely, the absence of psychological safety triggers the self-censoring instinct, which can also drive you into a state of high-context dependency.[67]

The psychological safety literature asks the critical question: what is the role of a leader? According to Bruce Daisley, the author of *The Joy of Work*, the role of a leader is to increase intellectual friction and, at the same time, decrease social friction. Creativity and innovation require collaboration, which means that ideas must be introduced counter to the status quo. Daisley calls this "constructive dissent" and "creative abrasion," which is achieved through intellectual friction. For intellectual conflict to occur, individuals must feel psychologically safe.[68]

Since human beings are behind ideas and decisions, especially those that run counter to the status quo, this may lead to social friction and defensiveness. When individuals become defensive, they experience the situation as a threat and rely on their assumptions and prior knowledge. They want to avoid pain and any perceived threat in their environment. If psychological safety is present, this will diminish some of that defensiveness and encourage ideas and innovation.

For example, if you know anything about working at Google, you understand that they operate under collectivism. They run on the team concept at every turn. Therefore, when they founded their project, Aristotle, the number one element was psychological safety. Google defined psychological safety as the following: "Team members feel safe to take risks and be vulnerable in front of one another." Can you imagine, in a team environment, if any member felt insecure or feared speaking up? Who would win in that scenario? No one, right.

Below are the stages of psychological safety:

- **Stage 1: Inclusion Safety**

---

[66] Clark, T. R. (2020). *The 4 Stages of Psychological Safety: Defining the Path to Inclusion and Innovation.* Berrett-Koehler Publishers.

[67] Allen, Kelly-Ann. (2020). *The Psychology of Belonging.* Routledge.

[68] Daisley, B. (2019). *The Joy of Work: The No. 1 Sunday Times Business Bestseller–30 Ways to Fix Your Work Culture and Fall in Love with Your Job Again.* Random House.

Inclusion safety means you feel safe being authentic, which means you are comfortable showcasing parts of your identity that are important to you. You are accepted not just for what you do, but who you are—your unique attributes and defining characteristics.

- **Stage 2: Learner Safety**
  Learner safety satisfies the basic need to grow cognitively. True learning can happen when you feel empowered to exchange ideas and ask meaningful questions. In addition, your feedback is valued, and mistakes are looked at as an opportunity to grow and learn.

- **Stage 3: Contributor Safety**
  Contributor safety satisfies the need to have an impact on your team or the organization. As a result, you feel a sense of safety in speaking up and using your skills, experience, and perspective to make a meaningful impact.

- **Stage 4: Challenger Safety**
  Challenger safety satisfies the need to improve one's environment by potentially bringing things up that run counter to the status quo. You feel a sense of safety when speaking up and challenging ideas because diversity of thought is valued, and as a result, one's social impact is minimal.[69]

So how does one build psychological safety within an organization? Here are some tips:

- **Encourage through behavior:** This relates to how a leader or others respond to bad news or dissension. Do they maintain poise, listen, and ask questions? Are they open to solutions and making others feel supported and safe?

- **Respond to vulnerability with care:** How do you respond to people's vulnerability? Showing the appropriate care, empathy, and compassion will send the message that being open and vulnerable is encouraged and supported. You tell others that it is a safe space to express themselves.

---

[69] Clark, T. R. (2020). *The 4 Stages of Psychological Safety: Defining the Path to Inclusion and Innovation.* Berrett-Koehler Publishers.

- **Avoid gossip:** How do you talk about employees and other people when they are not around? If you run others down or share harsh criticisms behind people's backs, you will destroy any trust between you and others. You will be considered someone unsafe to share the news with.

- **Allow people to talk about themselves:** When we show an interest in someone and ask them good and pertinent questions, the brain is most engaged. It builds trust and safety.

- **Emphasize the "why":** Invite different voices into a conversation. Constantly verbalize why diverse voices are essential to the organization and essential to innovation and creative thinking.

- **Be specific and "see" people:** It may be essential to avoid vague terms or phrases like "You are such a good worker," "I appreciate how you respect people," etc. It is vital to attach specific examples or specific character traits to those phrases so there is clarity, and the person knows you "see" them.

- **Allow people to name areas of growth:** It may be necessary for individuals to pinpoint specific areas they are working on so they are not hidden and can be encouraged. Also, making time for something like this may create the freedom to think creatively and innovatively. But, again, not everyone likes doing this, so it is critical to respect that by providing different avenues for individuals to share their voices in this way.

- **Communicate with individuals about how their input was used:** Asking for someone's feedback is positive. Still, the thing that builds trust and a feeling of psychological safety is telling the individual how their input was used and, if it was not used, why it may not be relevant to a particular situation or decision.

When we talk about Diversity, Equity, and Inclusion, we understand that diversity is a fact and inclusion is a choice. Diversity is a matter of composition. There is some type of diversity in any organization: cognitive diversity, intellectual diversity, and identity diversity. The question is, how well are you drawing it out, and how well are you creating space for it? Psychological safety and trust are a way to draw more of it out. If we do not allow differences to emerge, productivity may become stagnant, inactive, and there will be no harvest.

Diversity + Psychological Safety + Trust = Increased Inclusion (an individual's ability to participate, have a voice, contribute, and belong).[70]

Intentionality

Hall (1976) indicates that when there are identity differences, especially primary identity markers such as gender and race, there may be less understanding of an individual's experiences, values, and beliefs. As a result, communication needs to become more explicit and intentional, and a posture of learning and understanding is vital to feel secure. So, what exactly should someone be intentional about? Messaging, relationship context, spatial context, and time are critical factors in intentionality.[71, 72]

- **Messaging:** How someone conveys something is critical to creating strong intentionality. High-context-dependent individuals pay particular attention to common ground and an organization's values. Through its messaging, an organization creates a strong identity to which an individual can latch onto and be anchored.

- **Relationship context:** In a relationship, individuals may be concerned with saving face and potentially avoiding conflict. Therefore, they may be hyper-aware of behaving in a particular space as they are hyper-sensitive to where discussions occur. Being clear and explicit about expectations in specific spaces and situations may provide security and reduce anxiety and stress for high-context individuals.

- **Spatial Context:** Spatial context is another critical element to which some individuals may be sensitive. Factors like emotional distance, touching, communication gestures, and auditory space, like giving someone time to complete a thought or a sentence without interruptions, are critical for the high-context individual.

---

[70] Ferdman, B. M., & Deane, B. R. (Eds.). (2013). *Diversity at Work: The Practice of Inclusion* (Vol. 33). John Wiley & Sons.

[71] E. T. (1976). *Beyond Culture.* New York, NY: Random House.

[72] Chang, E. C., Downey, C. A., Hirsch, J. K., & Lin, N. J. (2016). *Positive Psychology in Racial and Ethnic Groups: Theory, Research, and Practice.* American Psychological Association.

- **Time:** Time is an aspect of intentionality. Taking the time to build authentic connections and relationships is very important. Taking the time to point out someone's value and contribution is essential. The currency of time communicates an important message; when you give someone your time, you directly tell them they are worthy, and you value them.

Summary

Here are four ways to mitigate high-context dependency in a group, and therefore increase a sense of belonging, and how to achieve them:

1) Build trust

- Recognize contributions.
- Challenge and support.
- Provide autonomy.
- Identify common ground.

2) Establish safety

- Encourage through behavior.
- Respond to vulnerability with care.
- Avoid gossip.
- Open spaces for personal sharing.
- Emphasize the "why."
- Acknowledge people's unique contributions.
- Promote a growth mindset and name areas of improvement.
- Tell people how their input was used.

3) Create a culture of intentionality

- Articulate values clearly in messaging.
- Provide relationship context and norms.
- Offer space for reflection and opinions and value them.
- Make time to build relationships.

Chapter 3

Bias

"I think unconscious bias is one of the hardest things to get at."
—Ruth Bader Ginsburg

Imagine this scene: You are at a soccer game one day, and you do not know many people in the crowd. You find a seat in the stands and sit next to a random person. You start conversing with the individual who tells you he is a pilot and loves soccer. You also encounter another individual who tells you they are a CEO of a large company and that they are passionate about soccer and have attended games for many years. A little hungry, you get up and go grab a bite to eat. While waiting for your food, you strike up a conversation with a friendly janitor.

So, after reading this story, our question to you is this: What was the gender, race, and ethnicity of the pilot, CEO, and janitor? We want you to be honest with yourself. There is no judgment, and only you will know your answers. However, hold those answers in the back of your mind as we discuss the barriers to belonging and biases. We will touch on how biases are formed, the role your brain plays, the potentially negative impacts, and inevitably how to alter them for more positive interactions.

What is Bias?

Bias can be defined as reaching conclusions toward or against someone in a preconceived, unreasoned, and assumptive way. Biases can be divided into **explicit bias** and **implicit bias.** The main feature that distinguishes explicit and implicit bias is your level of awareness.

Explicit Bias

Explicit bias occurs when you are conscious and aware that you have a bias, and you may deliberately act in a way that perpetuates it. For example, negative explicit biases can take the form of hateful speech or symbols directed towards a group of people that share an identity. Explicit bias must include an autobiographical knowledge of self (you know who you are and who you want to be) and a future state you want to achieve.[73] An unfortunate example of explicit bias occurred in 2022 when an individual opened fire in a supermarket in a predominantly Black neighborhood, shooting 13 people and killing 10, who were mostly Black, and had the goal of taking as many Black lives as he could. His White supremacist ideology revealed there was an awareness of who he was and an imagined future he wanted to achieve— a society free of people of color.

Perhaps you are in an organization and an individual is explaining they are wary of increasing the number of racially diverse individuals in the organization because he or she believes the standard of achievement will drop. Using the framework expressed above, the individual assumes that White people are more competent than non-Whites. There is also an imagined future the individual is trying to achieve, an organization that has more Whites than people of color because he or she believes racial composition is correlated to outcomes in the organization.

Implicit Bias

Implicit bias occurs outside of your conscious thoughts, and you may not be aware of them. Therefore, your actions are often automatic and not deliberate. An example could be feeling discomfort when someone who does not share your identity approaches you as you walk on a sidewalk.[74] You may not have a reason to fear the individual, but you have feelings of uneasiness. When thinking about implicit bias, it is important to remember that the brain

---

[73] Siegel, Daniel J. (1999). *The Developing Mind: How Relationships and the Brain Interact to Shape Who We Are.* Guilford Publications.

[74] Steele, Claude M. (2011) *Whistling Vivaldi: How Stereotypes Affect Us and What We Can Do.* WW Norton & Company.

is good at making mental associations. Imagine you are having a bad day and the mere sight of a trusted friend or colleague makes you feel better. You have made positive associations with that individual and seeing them automatically makes you feel a sense of comfort.

An important way to understand implicit bias is to understand memory— how we as human beings remember and recall things. First, one of the primary functions of memory is for the brain to remember past experiences, especially when there is heightened emotion, and apply it to make predictions about future events. When we experience things, the brain creates a series of neural networks that represent the experience we have. These networks store "information" about specific actions, people, groups of people, how you felt in the situation, your physical sensations, and the experiences of other people in that situation. So, if specific neural networks have been activated in the past, the probability of activating those same neural networks in the future increases, especially when you see and hear something that reminds you of a past event.[75] Have you ever had the experience of being with someone and specific memories are recalled from your life together? Maybe you have a relative that tells the same stories repeatedly when they see you. Their neural networks are activated when they are with you and your family, and you recall the associations you have with that individual.

It is important to note that these neural networks can form unconsciously when we are not aware of it. For example, at the time of writing this book, our son is two years old. Our son was born during the COVID-19 pandemic, and he was at home with us for a long time. When it was time to transition to daycare, it was a tough experience for him. Every time we would make the turn and pull into the parking lot of his daycare, he would make a heart-breaking sad face and start to cry. His brain had created neural networks that associated the daycare with negative emotions. This is significant because his higher order thinking abilities are not fully developed yet. That means that our son is not thinking, "Well, this may be a better day compared to yesterday, and maybe my parents forgot something so we will just be in and out." No. Our son has made associations with his daycare and recalled specific emotions in relation to the sight of the daycare. These implicit associations occur using the parts of our brain that develop at a young age, and these associations often do not require conscious processing, the higher order

---

[75] Siegel, Daniel J. (1999). *The Developing Mind: How Relationships and the Brain Interact to Shape Who We Are*. Guilford Publications.

thinking parts of our brain. The point here is that we are using the primitive and non-conscious parts of our brains to make these associations.

Over time, your mind and brain will forget the details, especially the specific circumstances and occasions when your values and beliefs were formed, and only remember that you have those values or beliefs. For example, you may think that people in the professional setting should not disclose information about their personal lives—they should not bring pictures of their families or partners because you believe there needs to be a separation between personal and professional. You may have forgotten the specific situations that led you to believe in this separation of work and home, but you believe it. It is part of how you view the workplace. Imagine you walk into work and see someone decorating their cubicle or office with personal items—you may react negatively because certain neural networks in the brain have been activated, particularly those that connect with your values around the separation of work and home.

Let us apply this framework to race. If from a young age we were exposed to a deficit view of a particular group of people, maybe from different adults, media, institutions, and society, your brain may have been internalizing these messages and forming neural networks relating to certain people or things without you being aware. If you are conducting an interview and an individual walks in with dreadlocks, walking or talking in a particular way, these associations may be brought to the surface and impact your decision making.

It is also important to note that the circumstances that lead to the activation of certain neural networks have a profound effect. For example, you may be stressed, overwhelmed, angry, anxious, or tired. In this state, when someone who does not share your identity makes a mistake or there is some breakdown in communication, your mind and brain may start to activate neural networks that have negative associations with the person's identity group. Due to the neural network activation, you may make assumptions about their competence to make sense of their behavior. Therefore, negative emotions can cause you to jump to unfortunate conclusions about a person.

Social psychologists divide **implicit bias** into **implicit prejudice**, which means having automatic negative feelings towards a particular social group compared to another, such as fear, anger, and disgust, and **implicit stereotyping**, which means assigning specific characteristics and holding certain beliefs about a social group, such as dangerous, lazy, unreliable, or incompetent.[76]

Now that you have some details about explicit and implicit bias, let us look deeper into the parts of the brain that are involved in biased interactions.

Bias and the Brain

Remember the story from a previous chapter where a manager asks each team member to think of a solution to a company problem and explain their answers in a short essay? After reviewing the team's work, the manager picks one paper out of the bunch and raves about the ideas. He then looks at the name on the paper and calls that person out. The person happens to be an African American man. When the manager sees who it is, he looks at the man with shock and a measure of doubt and asks, "You came up with this?" What might have been going on in the brain of the African American man when the manager asked that question?

Dr. Evian Gordon, a mental health and brain expert, describes the brain's threat and reward system as the "approach-avoid" response. In other words, when an individual steps into a new environment, the brain will immediately start to determine whether that environment is "good" or "bad." If it is deemed "good," the person will allow for engagement. If it is deemed "bad," the brain may signal for the individual to retreat. The mind is constantly playing a game of tug of war. It is continually assessing the benefits and risks of your environment. If the threat level outweighs the benefit, then the retreat signal will be turned on.[77]

Amygdala

The portion of your brain responsible for the "approach-avoid" process is called the limbic system—the emotional center, specifically the amygdala. The amygdala is instrumental in controlling the "approach" or "avoid" response. In fact, the limbic system can detect "good" and "bad" stimuli

---

[76] Malpas, J., "Donald Davidson", *The Stanford Encyclopedia of Philosophy* (Winter 2012 Edition), Edward N. Zalta (ed.).
https://plato.stanford.edu/archives/win2012/entries/davidson/.

[77] Gordon, E. (2000). *Integrative Neuroscience: Bringing Together Biological, Psychological and Clinical Models of the Human Brain.* CRC Press.

within a fifth of a second. Thus, the amygdala is like a watchdog that is always aware and sensitive to "approach" or "retreat."

For example, suppose an individual or a group has been continually exposed to environments of exclusion or unbelonging. In that case, the amygdala becomes a well-oiled machine and becomes more efficient at locating signs of unbelonging, even if this is not the intention.

To prove this theory, researcher Elizabeth Phelps and her colleagues performed a study using the following steps:

1. Participants were brought into a room and connected to instruments that measured perspiration, heart rate, blood pressure, respiration, and the release of stress hormones.

2. Researchers then showed the participants a blue light, and as expected, there were no atypical readings.

3. Then they again showed the participants a blue light but gave them a mild shock on the wrist. They did this same step several more times, conditioning the participants.

From that point onward, when the participants were shown the blue light without being shocked, researchers saw increased sweating, heart rate, blood pressure, respiration, and an increase in stress hormone release. In addition, when they assessed the brain activities of these participants during the experiment, they saw an increased activation of the brain in the amygdala. Thus, the investigation supports the understanding that the amygdala is involved in fear conditioning and creating emotional memories.[78]

Think about this in your own life. What do you fear? What causes you anxiety? Take a minute and reflect on the experiences you may have had that created those fears in your mind. For example, many Black employees and stakeholders explain that they fear having to constantly manage their hurt and pain following brutal acts of violence against people of color. They fear having to explain their emotions or having a leader, manager, or colleague make an insensitive comment about the event or situation.[79] For many

---

[78] Kubota, J. T., Banaji, M. R., & Phelps, E. A. (2012). "The Neuroscience of Race." *Nature Neuroscience*, 15(7), 940-948.

[79] Brown, Karen. (2021). "The Fear Black Employees Carry." *Harvard Business Review Digital Articles*, April, 1–8.
https://search.ebscohost.com/login.aspx?direct=true&AuthType=url,cookie,ip,custui d&custid=infohio&db=buh&AN=150220880&site=ehost-live&scope=site.

employees of color, they carry these moments with them, along with the fear of prejudice, being misunderstood, and the stress of having to manage their exhaustion.

Understand that the brain receives about eleven million pieces of information every second, but it is only equipped to process forty pieces. So, the brain must take cognitive shortcuts to make sense of the information it receives. The brain loves patterns and predictability because it can assess things faster when patterns are applied. Therefore, experiences, memories, and patterns play a massive role in how a brain deciphers good from bad, safety from danger, or certainty from uncertainty. Those assessments by the brain will create specific emotional reactions from an individual, such as fear.

Prefrontal Cortex

The amygdala assesses an environment for "good" and "bad," as we explained. The brain's prefrontal cortex deals with higher functions, such as reasoning, planning, focus, attention, inhibiting inappropriate behavior, impulse control, and delayed gratification. There is a negative correlation between how someone experiences threat activation and the resources available for the functioning of the prefrontal cortex. If the amygdala sends off warning signals, then one's ability to access those higher functions in the prefrontal cortex is diminished.[80] In other words, if one experiences fear then they may have a more difficult time acting rationally.

With the prefrontal cortex not functioning optimally, the amygdala and other components of the emotional brain are the primary drivers of behaviors. This means that emotions take over. When the amygdala is in the driver's seat, its ability to reason, process information, and break situations down rationally is compromised. As a result, people are more likely to be defensive and become hypersensitive to signs of rejection and unbelonging, such as not smiling, body language, and tone of voice. Minor stressors can become significant stressors when the amygdala hijacks behavior.[81]

What does this look like in real life? If individuals enter your organization and do not feel welcome, they are more likely to be disengaged, not take risks, and experience loneliness. Imagine that same new employee gets tasked with finding a solution to a company issue requiring them to work with others. If

---

[80] Goleman, D. (2005). *Emotional Intelligence.* Bantam.

[81] Goleman, D. (2005). *Emotional Intelligence.* Bantam.

that employee feels uncertain or anxious, their focus will be on their negative feelings, not the task. Their emotions (amygdala) will suppress the higher-order thinking (prefrontal cortex) and negatively impact their productivity.

Now, why are we repeating this information about the amygdala and the prefrontal cortex? Well, when the emotional brain (amygdala) is in the driver's seat, it is essential to recognize some of the following negative behaviors:

- **Overgeneralizing:** This is an all-or-nothing way of thinking. For example, imagine you are upset with a friend, partner, or spouse. During confrontations, you may use phrases like "you never" or "you always" despite there being times when they demonstrate the desired behavior.

- **Tunnel vision:** This entails focusing on the less important details of a situation. For example, imagine a supervisor who gives ten compliments and one piece of constructive feedback to facilitate an individual's growth. Tunnel vision will cause the recipient to deemphasize the ten positive comments and fixate on the one seemingly negative comment.

- **Jumping to conclusions:** This is the process of assuming. For example, suppose an employee misses an important deadline. The boss immediately thinks the error was caused by laziness. The boss made unfair assumptions and reacted with emotion rather than considering the circumstances that led to the particular outcome.

- **Magnifying a problem:** This entails making something small into a big deal. For example, if individuals receive constructive feedback, they interpret it as, "I might be fired," even though the feedback was minor and intended to help the individual grow.

- **Personalizing**: When individuals personalize something, they turn the blame inward, thinking or saying to themselves, "Everything is going wrong because of me!"[82]

How do we counteract the emotional brain taking control? As discussed in the previous chapter, scholar Barbara Fredrickson indicates that it takes

---

[82] Reivich, K., & Gillham, J. (2008). *Penn Resiliency Program Core Skills Manual.* Unpublished manuscript, University of Pennsylvania.

more positives to counteract a single negative in her "positivity ratio." So, if an individual perceives unbelonging, leaders and colleagues will have to work harder to earn back someone's trust.[83]

We suggest the leaders and co-workers will need to do the following to regain a person's trust:

- Listen without judgment
- Ask open-ended questions
- Use a tone that demonstrates empathy and concern for their well-being

Then the harmed individual can gain some trust and feel an improved sense of belonging. This will take time and repetition to become a conditioned response, as we saw with the experiment involving the blue light and shock at the start of this section.

Implicit Versus Explicit Bias and the Amygdala

Cognitive Neuroscientist Elizabeth Phelps and colleagues performed an experiment where they showed Black and White participants the faces of other Black and White college students with a neutral expression. The experiment measured implicit attitudes at the unconscious level and explicit attitudes using the IAT Implicit Association Test (a test developed by researchers at Harvard, UVA, and UW, more than two decades ago).

The experiment measured the startle response, an involuntary reaction, by measuring:

- The strength at which participants blinked
- The explicit—clearly stated—racial attitudes of the participants
- The blood oxygenation levels of participants in different parts of their brains

Below are the conclusions with regards to White individuals who had an increased amygdala activation, the brain's fight or flight response when fear or aggression are present:

- The White individuals had greater implicit (unconscious, implied) racial attitudes against Black faces showing an elevated startle response.

---

[83] Fredrickson, B. L. (2013). *Updated Thinking on Positivity Ratios.*

- There was no correlation between the amygdala and explicit (stated plainly) racial attitudes. In other words, they were not aware that they felt that way.

The conclusion is that individuals must pay attention to their racial attitudes, both implicit and explicit. Race does not produce bias, but it is the associations with a specific race that may cause an increase in the amygdala response. Both implicit and explicit biases cause people to project division and make others feel unwelcome and unsafe.[84]

It is easy for our negative implicit biases to go unchecked; negative implicit biases are associated with our brain waves. For example, right now, you are learning and actively engaging. This is called the beta state. However, you receive information bypassing your conscious mind into the unconscious arena in the alpha state. It is like when someone is watching TV and you try to talk to them. Unfortunately, they almost cannot hear you or receive anything you are saying because they are completely absorbed in the program.

Research shows that the more robust your alpha state, the more excellent the opportunity for internal biases. Therefore, we must be alert—self-aware of our feelings—and intentional about what we are consuming and how we are acting.[85] For leaders who hope to counteract biases in the alpha state, it is important to humanize the people you lead and serve. Leaders can humanize their employees by:

- **Learning names:** A person's name is intimately tied to their identity. Pay particular attention when someone is telling you their name and keep using it in sentences. This activates the reward pathways of the brain. It builds connection.

- **SOFTENing your body language:** Research shows that two-thirds of communication is body language. So, it is essential to S-O-F-T-E-N your body language. The acronym stands for Smile, Open

---

[84] Phelps, E. A., O'Connor, K. J., Cunningham, W. A., Funayama, E. S., Gatenby, J. C., Gore, J. C., & Banaji, M. R. (2000). "Performance on Indirect Measures of Race Evaluation Predicts Amygdala Activation." *Journal of Cognitive Neuroscience*, 12(5): 729-738.

[85] Grabot, L., & Kayser, C. (2020). "Alpha Activity Reflects the Magnitude of an Individual Bias in Human Perception." *Journal of Neuroscience*, 40(17): 3443-3454.

Posture, Forward Lean, Touch (e.g., elbow bump, shaking hands), Eye Contact, and Nod.[86]

- **Add personal context:** Make sure to introduce yourself and your role and add context. Let them know what is involved on your part. In other words, give your title meaning. Introduce them to someone else and let them know something about that person.

- **Ask open-ended questions:** Ask open-ended questions that allow for the individual to elaborate. Do not use close-ended questions that can be answered with a simple "yes" or "no."

Many of you say, "I get all this, but how do I control an involuntary reaction?" You can control your implicit responses by changing or redirecting your anchor biases, which we will address later in this chapter.

Individual Bias

Researchers surveyed 1,918 workers from large companies with 1,000 or more employees to assess negative biases in their perceptions. To start, the researchers defined bias as an employee having a positive self-rating but perceiving their supervisors' rating as unfavorable. The researchers discovered that one in ten workers generally perceive negative biases in their workplace. In addition, these workers reported having higher levels of negative emotions, burnout, disengagement, loneliness, isolation, and a higher likelihood of leaving their companies. The study also noted that people of color are more likely to perceive negative bias at work than White individuals.[87]

When an individual has a negative bias, it can be described as reaching conclusions against someone in a preconceived, unreasoned, and assumptive way.[88] Suppose, like with the employees in the study above, someone who believes they are superior at work is experiencing a negative bias toward them. In that case, they may feel as though the person has misjudged them

---

[86] Gabor, Don. (2011). *How to Start a Conversation and Make Friends: Revised and Updated.* Simon and Schuster.

[87] Hewlett, S. A., Rashid, R., & Sherbin, L. (2017). "When Employees Think the Boss is Unfair, They're More Likely to Disengage and Leave." *Harvard Business Review Digital Articles*, 2.

[88] Merriam-Webster. (n.d.). "bias." *Merriam-Webster.com Dictionary.* January 4, 2020, https://www.merriam-webster.com/dictionary/bias.

based on an unfounded belief. This perception will directly impact their performance.

For example, suppose your department manager constantly micromanages your day, but not your peers. You perceive this boss as having a negative bias against you because you are a minority, unlike your peers. Will you be incentivized to perform at your highest level? To some degree, the honest answer is that you will most likely act in a diminished capacity. Conversely, if you believe your superior fully supports you, your focus and productivity will reflect those positive emotions.

One Yale Child Study Center conducted an experiment with 135 educators. These teachers were instructed to watch short videos of White and Black students and were told that the students may or may not be engaged in negative behaviors. However, the students did not participate in any negative behaviors by design. As we said, the teachers did not know this. The study revealed that Black and White educators tracked kids of color more closely than White students. What does that show? Many individuals harbor preconceived assumptions about others, and they may or may not even realize it.[89]

So, where do individual biases come from? They come from reference points or anchors. Direct and indirect experiences shape our beliefs, such as personal interactions, television, and social media. These anchor points are stored in the unconscious mind. However, before delving into the unconscious mind, we must discuss three essential aspects of the mind.

There are three main aspects of the mind: the conscious mind, the preconscious mind, and the unconscious mind. To illustrate these essential parts, look at the iceberg, then read the following descriptions below:

---

[89] Gilliam, W. S., Maupin, A. N., Reyes, C. R., Accavitti, M., & Shic, F. (2016). "Do Early Educators' Implicit Biases Regarding Aex and Race Relate to Behavior Expectations and Recommendations of Preschool Expulsions and Suspensions?" *Yale University Child Study Center*, 9(28): 1-16.

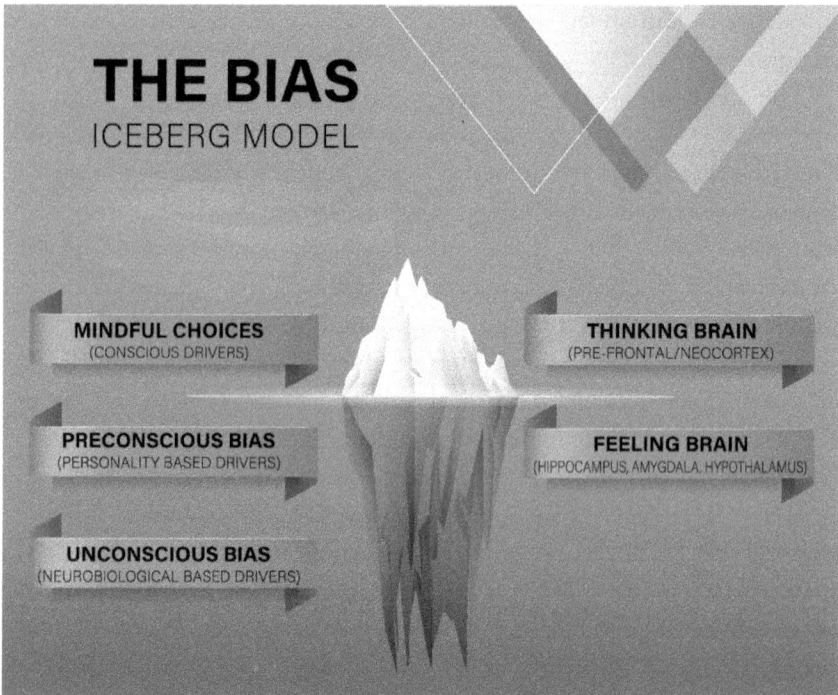

- **Conscious:** The visible part of the iceberg above the water is like the conscious mind. These are things a person is aware of at any given moment: thoughts, feelings, memories, and emotions. Individuals can easily talk about their conscious mind's thoughts, feelings, and emotions because they feel them in real-time.

- **Preconscious:** The portion of the iceberg just beneath the surface is called the preconscious mind. This is often referred to as "available memory." For example, if you were asked to give someone's phone number or address, that information may easily be accessible. Even though you were not consciously thinking about the phone number, you can easily recall it. When you remember a phone number or address, your preconscious thoughts move to the conscious mind.

- **Unconscious:** The iceberg's deepest parts are analogous to the unconscious mind. This is the warehouse of all memories, motives, values, worldviews, and beliefs. Here is the thing—the unconscious mind usually dictates a person's reactions and helps individuals make sense of the world.[90, 91, 92, 93]

From which part of our mind do we get our anchors or reference points about groups of people? We recall them from the unconscious mind. We form anchor points and store them in the unconscious mind through our socialization and the different things that give meaning to our worlds throughout our lives. These influences include media, friends, family, religion, work, school, and even the toys we play with as children.

Let us look at the stages in life's experiences where individuals obtain values, beliefs, and attitudes and create those anchor points:

- **The beginning:** At birth, individuals are born into a society with specific values, beliefs, or perspectives about various dimensions of diversity.

- **First socialization:** Children learn values, beliefs, and norms from parents, relatives, teachers, and people they love and trust. They start to model behavior, set expectations and ways of being, and shape what they value and believe as they view the world around them.

- **Institutional and cultural socialization:** As individuals grow older and enter institutions such as schools and places of worship, these environments often reinforce previous exposures during their first socialization. During this time, individuals start to gain a sense of their identities and the identities of others.

- **Enforcements:** Individuals get rewarded for actions that fit into the expected behaviors and values of communities and institutions.

---

[90] Boag S. "Conscious, Preconscious, and Unconscious." *Encyclopedia of Personality and Individual Differences.*

[91] Cherry, K. (2020, December 9). "The Structure and Levels of the Mind According to Freud." *Verywell Mind.* https://www.verywellmind.com/the-conscious-and-unconscious-mind-2795946.

[92] Walinga, J. (2014, October 17). "2.2 Psychodynamic Psychology." *Introduction to Psychology,* 1st Canadian Edition. https://opentextbc.ca/introductiontopsychology/chapter/2-2-psychodynamic-and-behavioural-psychology/.

[93] Mcleod, S. (2015, January 1). "Freud and the Unconscious Mind." *Unconscious Mind: Simply Psychology.* January 1, 2015. https://www.simplypsychology.org/unconscious-mind.html.

Deviations from those behaviors or values may result in punishment.

- **Results:** This is when individuals establish and embody the beliefs and values they have learned and experienced.[94]

- **Actions:** The actions have two potential outcomes: individuals either start to promote what they have learned or question it.

Anchor Bias

Let us say you walk into a store and see this.

If you have been conditioned to be price-conscious, which one would you most likely buy? Yes, the one with the $300 crossed out. Why? Because it is on sale! You may not know a lot about the item you are buying, like the quality of the material or the stitch count. But you rely on limited information to make your decision. What drives an individual's propensity to react this

---

[94] Harro, B. (2000). "The Cycle of Socialization." In M. Adams, W. J. Blumenfeld, R. Castañeda, H. W. Hackman, M. L. Peters, & X. Zúñiga (Eds.), *Readings for Diversity and Social Justice: An Anthology on Racism, Antisemitism, Sexism, Heterosexism, Ableism and Classism.* New York: Routledge.

way? It is your anchor point: being preconditioned to believe the lower cost is best. It does not make your decision to buy the lower-priced item more correct.

Therefore, it is essential to take the time to make more informed decisions and not just rely on your anchor biases. Being more aware of your unconscious biases will allow you to make a more informed and controlled decision, thus expanding your anchor points.

Affinity Bias

Author Claude Steele described a friend's situation when he was a graduate student. His friend was an African American man named Brent Staples who attended the University of Chicago. Brent noticed that when he would walk down the street in the presence of Whites, they would avoid him or cross the street. So, he decided to perform an experiment where he would walk while whistling Vivaldi. He discovered that Whites avoided him and crossed the street less when he whistled classical music.[95]

Brent's experience represents a bias called a similarity or affinity bias. Individuals are biased towards people who look like them or share the same identity or norms. Essentially, people feel more comfortable with others like themselves. Does it mean individuals are "bad" for possessing these affinity biases? Well, no, but individuals must learn how to be aware of these biases and counteract them, especially when they negatively affect other individuals.

The brain loves to use patterns for quick reviews when doing assessments. So, people who look similar to us or share similar identities or norms are labeled "safe" by the brain. However, different individuals can cause the brain to signal fear. Have you ever walked down the street and seen someone who does not fit in with the surroundings? Did you feel more alert to this person's actions? What about the "stranger danger" concept we were taught as kids? From an early age, children are taught to avoid the unfamiliar, the unknown. That belief becomes an issue when it is carried over into bias against a particular racial or ethnic group and causes discrimination or avoidance.

However, affinity biases go beyond just the brain deciphering "good" or "bad" based on someone's race. It also encompasses an individual's ability to

---

[95] Steele, C. M. (2011). *Whistling Vivaldi: How Stereotypes Affect Us and What We Can Do*. WW Norton & Company.

see distinguishable differences between people of the same race. For example, Crystal and Pascal, both Black, attended church one day. They had attended this church for some time and knew many of the parishioners. A White woman approached them excitedly as they walked in and asked about their baby. They had not had a baby at that time, but the only other Black couple in the church had just had a child. A case of mistaken identity? Yes. Was this woman's intent to be offensive? No, of course not, but it does demonstrate this concept of the limitations of one's ability to see differences between others of another race.

To further illustrate the trouble with facial recognition across races, let us look at another study. In this study, subjects were placed into two groups based on phony personality tests to give them a sense of group identity. The researchers showed the participants faces of people they explained were part of their personality group (the "ingroup") and faces of people that were not in their personality group (the "outgroup"). The results showed that the subjects paid closer attention to ingroup faces and remembered them more accurately compared to outgroup faces. Therefore, the researchers suggest that those who share group membership of some kind, whether its personality or identity, tend to have a memory advantage when it comes to remembering and recognizing faces, thereby underscoring an affinity bias.[96]

As an educator, Pascal saw the effect of this. He taught a racially diverse group of students, and the group size was substantial. Yet, he experienced more difficulty distinguishing between the White and Asian students than the Black students.

So, how can this negatively impact individuals? Are these not just innocent mistakes? Yes and no. For example, let us look at The Innocence Project, a non-profit group that uses DNA to free wrongly accused individuals serving time for a crime they did not commit. Many of them were convicted because of eyewitness accounts that resulted in misidentification. The Innocence Project (2014) reported that they have helped exonerate 325 people. Seventy-two percent of these individuals were convicted because of mistaken eyewitness identification. Of those misidentifications, fifty-three percent involved a White witness misidentifying a Black person.[97] This shows

---

[96] Herzmann, G., & Curran, T. (2013). "Neural Correlates of the In-group Memory Advantage on the Encoding and Recognition of Faces." *PloS one*, 8(12): e82797.

[97] "Eyewitness Identification Reform." (n.d.). Innocence Project. Retrieved December 3, 2021. https://innocenceproject.org/eyewitness-identification-reform/.

that while it may be an unconscious mistake, these mistakes can have negative consequences for individuals if left unchecked.

The Importance of Diversity in Our Social Networks

So, when we think about affinity bias, we need to look at a study completed by the Public Religion Research Institute (PRRI). The PRRI designed a survey to understand the diversity of American social networks.[98] The survey concluded the following:

WHITES:

- Studies show that among White Americans, seventy-five percent of people disclosed "important matters" to those who share their racial identity.
- Fifteen percent indicated social networks were racially mixed.

BLACKS:

- Among Black Americans, sixty-five percent of people disclosed "important matters" to those who share their racial identity.
- Twenty-three percent of Black Americans' social networks were racially mixed.

LATINX:

- Among Latinx Americans, forty-six percent of Latinx Americans disclosed "important matters" to those who share their racial identity.
- Thirty-four percent of Latinx Americans' social networks were racially mixed.

The statistical results clearly indicate a comfort level with others of ingroup over outgroup.

---

[98] Cox, Daniel, Juhem Navarro-Rivera, and Robert P. Jones. "Economic Insecurity, Rising Inequality, and Doubts about the Future." (2016, August 1). Public Religion Research Institute (PRRI). https://www.prri.org/research/survey-economic-insecurity-rising-inequality-and-doubts-about-the-future-findings-from-the-2014-american-values-survey/.

Along the same lines, researchers gave a survey to 35,000 people who had connections, in the form of friendships and relationships, with people who did not share their racial identities. The researchers identified the racial diversity in the participants' states and counties. The results showed that for individuals who reported "above average contact," which was defined as having two or more relationships with people who did not share the participant's racial identity, their attitudes towards racial outgroups were not dependent on the diversity within their states or counties. Rather, their relationships dictated their attitudes. Consequently, for those individuals that had "zero interracial contact," the more diverse their states were, the more negative their attitudes toward people that did not share their racial identity.[99]

What does this mean? Close connections and relationships with those of a different race or ethnic group can help counteract negative biases.

Negative Aspect of Bias and How to Change

We have talked about how biases are formed and how attitudes toward others are implicit and explicit. Implicit attitudes are formed at the unconscious level, meaning they are involuntary and generally unknown. However, straightforward attitudes are conscious, deliberately created, easier to self-report, and easier to change.

When you have a negative bias, the brain does three things, as we discussed:

- **Recognize a difference:** The brain recognizes a difference and then reaches deep down into the unconscious mind to access a reference point for that difference. In other words, it asks where have you seen someone like this before.

- **Anchor:** The brain creates an association or stereotype of an individual or group which becomes the anchor in the subconscious mind. A stereotype can be defined as an oversimplified impression, image, or truth about an individual or a group of individuals.

---

[99] Redford, L. (2018). "Personal Relationships Weaken Interracial Bias." *Society for Personality and Social Psychology: Character & Context*. https://spsp.org/news-center/blog/relationships-weaken-bias.

- **Negative Association:** You have negative associations with other individuals or groups. Implicit bias often shows up subtly; it is not overt, so you may not be aware of it.

When you understand these facts, you can better understand how to change your anchor biases. We must create an optimistic view for change to occur using the following points:

- There should be no shame or guilt assumed by anyone. We must be willing to extend grace, educate, and encourage individuals to counteract these biases.

- Having negative implicit biases does not make you a bad person, but *you are responsible for recognizing them*. It is also important to recognize that counteracting implicit biases is possible.

  - Charlesworth and colleagues conducted 4.4 million tests of implicit and explicit attitudes towards different groups based on age, ability, body weight, race, skin tone, and sexuality. Over the course of thirteen years from 2004-2016, the researchers saw a reduction in explicit and implicit biases, especially when it came to sexual orientation, race, and skin-tone.[100] When it came to race and skin tone, millennials showed the most significant decrease in implicit biases.

    While we can look positively at this research, it is important to recognize how people view progress. Some people look at progress as the past till now and some look at progress from where we are to where we want to get to. Even though implicit racial attitudes are possibly becoming more positive, a Pew research study showed that about six in ten Americans believe that race relations in the US are generally bad.[101] In other words, we have more work to do, and that

---

[100] Charlesworth, Tessa ES, and Mahzarin R. Banaji. (2019). "Patterns of implicit and explicit attitudes: I. Long-term change and stability from 2007 to 2016." *Psychological Science* 30(2): 174-192.

[101] Horowitz, Juliana Menasce, Anna Brown, and Kiana Cox. (2021, September 22) "Race in America 2019." Pew Research Center's Social & Demographic Trends Project. Pew Research Center.

is why obtaining the skills and tools to bring unity, reconciliation, and belonging are important.

- It is essential to think about implicit biases' impact on an individual. Individuals subjected to actions resulting from negative implicit biases report a lower level of well-being. So, your words and actions can harm another, even if unintended. Your goal should be to create an environment of safety and inclusion for others.

- Flood your mind with positive images of different groups of people. You can also do this by getting to know people who do not share your identity or norms. Basically, create new anchor points to counteract the old. The goal should be to develop more positive associations.[102]

- To drive the previous point home, let us look at some research. Forscher and colleagues were very interested in this topic of bias and did a meta-analysis of 492 studies (with over 80,000 participants) to investigate the effectiveness of implicit bias interventions.[103] The authors found that interventions that targeted the associations people made with groups of individuals, both implicitly and explicitly, were most effective. Examples of effective interventions included individuals setting goals and organizations articulating clear standards for counteracting bias. Those interventions that resulted in consequences, as well as those that involved affirming people for demonstrating anti-bias behavior, were not as effective.

Therefore, if you are a leader hoping to decrease negative biases in the workplace, the research above illustrates that:

- You can **target people's associations** by exposing yourself and those you lead to counter stereotype examples. You can do so by having positive images on your walls or purposefully using

---

[102] Ross, H. (2008). "Proven strategies for addressing unconscious bias in the workplace." CDO Insights, 2 (5). Washington, DC: Diversity Best Practices. Retrieved on July 10, 2016.

[103] Forscher, P. S., Lai, C. K., Axt, J. R., Ebersole, C. R., Herman, M., Devine, P. G., & Nosek, B. A. (2019). "A meta-analysis of procedures to change implicit measures." *Journal of Personality and Social Psychology*, 117(3): 522.

examples, images, and quotes from people of color who have contributed to particular fields.

- You can practice **perspective taking** by looking at situations from another's point of view. For example, there was an organization that gave an anonymous survey to its stakeholders. They recognized that different groups were being impacted by bias, so they posted the comments people wrote in public areas for stakeholders to see and read. Stakeholders within the organization were able to read stories of how bias had adversely impacted their peers.[104]

- **Set personal organizational goals for counteracting bias**. Having clear goals are effective for people to hold themselves and each other accountable to their goals. When you set goals, the brain starts to monitor and become sensitive to your behaviors in relation to those goals.

- **Having relationships with people that do not share your identity** enables you to understand them as individuals, limits assumptions, and continually updates your reference points.

Interventions that are useful, but not as effective, according to the study above include:

- **Having negative consequences for behaviors.** Therefore, actions like making anti-bias trainings mandatory were not as effective at counteracting negative biases.

- **Giving participants feedback** that they are unbiased and have good morality and competence was not as effective because it did not address the root of the negative biases.

Also, note that you can take the Implicit Association Test (IAT) mentioned earlier. This resource has helped millions of people identify their implicit and unconscious biases. You can go to the website https://implicit.harvard.edu/implicit/takeatest.html to take the IAT and identify any unconscious biases for free.

---

[104] Kepinski, L & Nielsen, T. (2020). *Inclusion Nudges Guidebook.* Independently published.

Understand that having negative thoughts does not mean you are a terrible person. Our dominant culture is set up to emphasize binaries, for example good versus evil. We get it—you want to be a good person. However, be slow to place value judgments on yourself and give yourself grace. Research shows that if you exercise self-compassion, you will continue to be motivated to move forward.[105] So, keep track of behavior, strive to improve, and know that sometimes your reactions are unconscious but can be altered if you want to purposefully create more positive associations.

Emotional Intelligence

Another way to be purposeful in interacting with others is to increase your emotional intelligence. Emotional intelligence directly relates to how you use your emotions. Improving your emotional intelligence involves the following:

- **Self-Awareness:** Know your emotions and recognize your feelings. This enables you to engage your neo-cortex and conscious mind and deactivate the emotional brain. As you become more self-aware, you become alert to the experiences as they are happening. When you become more self-aware you make the implicit become more explicit and can change your behavior. However, self-awareness must be coupled with intention. Intention means that you have purposefully set goals for a specific behavior and are committed to achieving them. If you are a leader and perceive implicit prejudice or implicit stereotypes, you may recognize them and set goals so you can reframe your thinking and impact your actions.

- **Manage the Emotions You Feel:** Engage the prefrontal cortex by asking yourself questions: Is this emotion appropriate? Asking questions is very powerful because it engages your higher order thinking and deactivates your emotional response. When you engage your higher-order thinking, you are able to take better charge of your emotions. Imagine you are in a meeting where you are stressed, anxious,

---

[105] Fernandez, R & Stern, S. (2020). "Self-Compassion Will Make You a Better Leader." *Harvard Business Review.* https://hbr.org/2020/11/self-compassion-will-make-you-a-better-leader.

and overwhelmed. Someone makes a comment that you consider careless or irrelevant to the discussion, and you start to get aggravated. You may notice that you are making assumptions about that person and ask yourself, "Are these emotions appropriate?" You may find that your emotions have nothing to do with the person but are instead from other situations that are contributing to your negative emotions. Considering an alternative explanation in the moment can cause your potential implicit prejudice or implicit stereotype to lessen.

- **Be an Optimist:** You need to recognize that change is possible. Optimistic thinking asks the question, "What could go right?" as opposed to our common default question, "What could go wrong?" Your mindset is important when it comes to bias. Optimistic thinking presents a buffer against shame, apathy, and hopelessness, feelings that discourage you from continually striving for unbiased behaviors.[106] For example, imagine you are in a meeting and say something that the people you lead consider biased. You are confronted by them and listen well to what they have to say. You recognize that if you continue to strive to be better, you can position yourself to gain trust and model behavior for those you are leading.

- **Empathy:** One of the things that sets the human species apart from other species is the size of our frontal lobe. We have the capacity to "perspective-take" and step into someone else's shoes as they are relaying their experiences. Research indicates that when we listen well and "perspective-take," our capacity for empathy increases.[107] Empathy humanizes others and allows us to appreciate the unique nature of people. Imagine that you are conducting a job interview and the candidate shows up late. The person you are interviewing is younger than many people in the office and has an underrepresented identity. You feel yourself jumping to conclusions about the individual, especially based on age. You can tell the individual is flustered, so you choose to

---

[106] Seligman, Martin EP. (2006). *Learned Optimism: How to Change Your Mind and Your Life.* Vintage.

[107] Zaki, Jamil. (2019). *The War For Kindness: Building Empathy in a Fractured World.* Crown.

take the time to understand why they are late. You find out that they are late because they were involved in a minor car accident. You then acknowledge that their encounter can be very stressful. Your ability to feel what they are feeling and acknowledge the challenges of the situation mitigates your bias and can also make the person feel more comfortable.

Summary

Here are ways to reduce bias:

1) Disarm the brain's fear mechanism by:
   - Building rapport
   - Making physical contact
   - Validating others
   - Engaging intentionally
   - Being positive

2) Counteract negative biases by:
   - Targeting people's associations by countering stereotypes
   - Practicing perspective-taking
   - Setting clear personal and organizational goals and accountability practices
   - Relating to others outside your identity group

3) Humanize people by:
   - Developing an awareness of your emotions
   - Managing your emotions
   - Being optimistic
   - Listening and empathizing

Chapter 4

Organizational Bias

"When we listen and celebrate what is both common and different, we become wiser, more inclusive, and better as an organization."
—Pat Wadors

Some years back, while participating on a hiring committee for his organization, Pascal and other committee members narrowed a list of candidates to a select few. As they began to review the resumes of this final group, they found one of the candidates had remained in her role for over twenty years. Given the current environment where a change in and amongst jobs is commonplace and most individuals strive to be promoted, this fact stood out as odd. One member of the panel spoke up and was quick to suggest that they dismiss this candidate solely based on this information, assuming it revealed something negative about her character.

At that moment, a female member spoke up. "Wait. I noticed this on her résumé, and during our interview, I asked her about it." That member went on to explain that the interviewee remained on the job for twenty years due to her family situation; her husband was pursuing a doctorate, and the position was conducive to their family's lifestyle. So, this explanation made sense, but it was something the others on the panel, including Pascal, had not considered.

What does this show? Well, for starters, it illustrates a scenario we call "fundamental attribution bias." When we do not understand specific facts, we will likely assign a biased judgment on a person's character. Second, it demonstrates that another's perspective opened everyone's eyes to a plausible reason that took the negative and turned it into something they could comprehend. Therefore, they did not dismiss this candidate based on a preconceived notion with the additional information. Instead, they judged her on her skillset and saw her as a favorable candidate for the job. The result: a more informed and unbiased decision was made.

Connection Between Individual and Organizational Bias

Organizational bias results from individual preferences that have been intentionally or unintentionally institutionalized and embedded into systems and structures within an organization. Personal biases must be embodied by large numbers or influential individuals for them to show up at the level of the organization. The hallmark of institutional bias is it not only impacts an individual but specific groups of individuals. This means some people in the organization will feel like they do not belong and may feel devalued. When that occurs, how can productivity happen? Does the organization inevitably suffer? Yes, of course.

We should explore some of the effects biases have on people, the impact on organizations, the society the organization is in, and the organization's productivity. First, since individuals are an organization's most important asset, how do biases impact employees, customers, or students when doing day-to-day activities?

To understand what we are referring to, we will look at Deloitte's 2019 State of Inclusion report. This survey reached out to 3,000 individuals in the United States from organizations with 1,000 or more employees and gave us a direct insight into how employees experience bias:

- Sixty-eight percent reported that bias harmed their productivity.
- Seventy percent believed bias had negatively impacted their engagement at work.
- Eighty-four percent said that bias negatively affected their happiness, confidence, and well-being.[108]

---

[108] Deloitte. (2019). *2019 State of Inclusion Survey Results.*
https://www2.deloitte.com/us/en/pages/about-deloitte/articles/unconscious-bias-workplace-statistics.html.

The results clearly demonstrate that organizational bias can lead to uncertainty and a lack of inner trust. When there is a lack of trust or confidence, individuals may not be as motivated to invest in the organization and may even be disengaged or withhold their ideas. Consequently, productivity is diminished.

In another such study, the consulting group McKinsey found that applying a rigorous process to decision-making was a one-to-one correlation with positive productivity. They happened to look at a German company that invested heavily in the market with minimal returns. Discouraged by its investments, the company took a step back and examined its decision-making processes.

The review of their practices revealed the following:

- The company discovered their investment decisions were always made by their high-level executives.
- The executives tended to agree across the board on most aspects, indicating they held similar biases.

Given that the executives' decisions had negatively impacted the bottom line, a new process around investment decisions was created by inserting a "devil's advocate," an individual who would:

- Challenge the thinking
- Open the views
- Break down some biases that direct decisions. This allowed for a new system that would stand whether the individual players remained or not.[109]

Organizational biases also have an impact on the society they are in. For example, the Bureau of Labor Statistics looked at the highest occupational levels within professional organizations and found that Latinx and Black Americans made up the lowest percentage of these professionals. In addition, a 2019 study by Crist Kolder Associates looked at CEOs of Fortune 500 and S&P companies. Only 8.7 percent of the 675 companies they reviewed had CEOs of color.[110]

---

[109] McKinsey Quarterly. (2017). "A Case Study in Combating Bias." https://www.mckinsey.com/business-functions/people-and-organizational-performance/our-insights/a-case-study-in-combating-bias.

[110] Gal, S., Kiersz, A., Mark, M., Su.R., & Ward M. (2020). "26 Simple Charts to Show Friends and Family Who Aren't Convinced Racism is Still a Problem in America."

Various commentators on this data partially attribute the underrepresentation of people of color to systemic or institutional bias in the business world. Systemic bias is when a particular identity group maintains an advantage over other identity groups. These advantages may result from individual or interpersonal biases embedded within various structures within an organization. Basically, the cumulative effect of individual bias can be seen at the level of the organization and greater society.[111]

Organizational Bias and the Hiring Process

If organizations do not employ standard processes to counteract bias, individual preconceptions can alter the organizational environment and decisions. So, imagine individuals belonging to or working for an organization with no procedures in critical decision-making, such as the hiring process. In that case, the choices about who to hire would most likely reflect the recruiting employees' beliefs or biases.

For example, at the University of Toronto, Sonia Kang conducted a study to create resumes for Black and Asian applicants. The participants applied for 1,600 entry-level jobs. Some of the submitted applications contained information pointing to the applicants' minority status, while the others deemphasized racial clues.[112]

For example, some applicants chose the name "Lamar" because it sounded distinctively African American but used the last name "Smith" because, according to the US Census Bureau, it is a common name amongst both Blacks and Whites. So, when a Black applicant's name was "whitened," it showed up as L. James Smith opposed to Lamar J. Smith. In addition, racial cues such as "Vice President of Aspiring African American Business Leaders" were used to describe activities for "unwhitened" resumes but were removed from "whitened resumes." The investigators then created email

Business Insider. February 1, 2021. https://www.businessinsider.com/us-systemic-racism-in-charts-graphs-data-2020-6.

[111] Jana, T., & Diaz Mejias, A. (2018). *Erasing Institutional Bias: How to Create Systemic Change For Organizational Inclusion. First edition.* Berrett-Koehler Publishers.

[112] Kang, S. K., DeCelles, K. A., Tilcsik, A., & Jun, S. (2016). "Whitened Résumés: Race and Self-presentation in the Labor Market." *Administrative Science Quarterly, 61*(3): 469-502.

addresses and phone numbers to determine how many applicants were invited for an interview.

The results showed that whitened resumes produced more job callbacks than African Americans and Asian resumes.

A follow-up experiment was done by Harvard University. With fifty-nine participants in the study, thirty-six percent of them "whitened" their resumes, and about a third knew of other individuals who had commonly practiced whitening a resume. Below are how these participants made their resumes more "White":

- **Changed the presentation of their names:** Half of the participants changed the appearance of their names. For example, many Asian participants adopted a common American first name while keeping their Asian last name.

- **Changed the presentation of their experiences:** The experiment participants "toned down" their experiences; they removed any activities that looked too ethnic or controversial.

- **Changed the description of their experience:** The students altered how they described their experiences.

- **Adding "White" experience:** College students added typically "White activities" to "blend in."

To fully understand this overall concept of organizational bias in respect to recruitment, we will explore the specific types of biases that frequently exist within an organization:

- **Affinity Bias:** Affinity bias occurs when we subconsciously feel a connection with someone because they have the same interests or a similar background, or it is when a person reminds us of someone we are familiar with.

  For example, Pascal was conducting an interview with an individual he had known for many years. He felt a deep connection with this individual and found himself asking questions and saying things to encourage them, even more so than he would do with someone he was not as familiar with. As he reflected on the interviews, he realized that he offered the familiar candidate more micro affirmations such as warmth, smiles, and positive body language, compared to the other candidates.

- **Confirmation Bias:** This is a bias that many individuals in organizations need to be wary of, especially in the hiring process. When we fall victim to confirmation bias, we tend to have a preconceived opinion of someone based on the college they attended, the family they are connected to, or the organization they worked for previously. We then subconsciously look for evidence to back up the opinions and perspectives we have toward someone.

  For example, Pascal recalls speaking with a client about confirmation bias, and the owner of the business recounted a situation where they were hiring an individual from a competitor. The competing organization had a negative reputation for having a toxic workplace culture, "stuck-up" employees, and individuals that were only concerned about the bottom line. The business owner had to check himself from looking for evidence confirming these negative characteristics in the candidate he was interviewing.

- **Attribution Bias:** This occurs when individuals attribute behaviors to someone's character rather than inspecting the circumstances and situations that resulted in a specific behavior. Stephen Covey says that we often judge ourselves by our intentions and circumstances but quickly attribute behaviors to a person's character or personality.[113]

  In Pascal's early twenties, he worked for a valet parking company at an upscale restaurant. He remembers driving to his first shift. He did everything right—he left an hour early to get to his destination about thirty minutes ahead of time. What he did not account for was rush hour traffic. He sat in bumper-to-bumper traffic in frustration and watched the clock tick past the start time for his shift. He was angry, mortified, and scared and thought he would lose a job he had not even started. He was afraid that his supervisor would attribute his tardiness to his personality and character rather than understanding the circumstances of his situation.

- **Conformity Bias:** This is where we deemphasize parts of who we are to fit in and feel a sense of belonging.

---

[113] Covey, Stephen R., and Rebecca R. Merrill. *The Speed of Trust: The One Thing That Changes Everything.* Simon and Schuster, 2006.

For example, if you are working among others that are less educated than yourself, you may hide or downplay your academic achievements to "get along" or "fit in" with them. Additionally, you may have aspects of your identity that are important to you, but you de-emphasize them because you feel as though you may not be accepted.

An example of a person's concern about conformity bias is when Pascal was conducting an interview for a prominent position at an organization. As Pascal spoke to a short-listed candidate, she revealed that she was in a same-sex marriage. She asked him about the DEI programs at the organization and how accepting the culture was to individuals that did not share the dominant sexual orientation. She indicated that she was actively trying to counteract conformity bias where people existed in a culture where they had to hide important aspects of their identities.

- **The Halo Effect:** The Halo Effect occurs when individuals have an aspect of their personality or accomplishments that you admire. These attributes alter the way you see them and overshadow the negative aspects they demonstrate.

During his time in education, Pascal remembers interviewing a teacher who had multiple accolades for research and academic ability. As he spoke with the candidate and listened to people's feedback, it was apparent that this individual would clearly struggle to connect with the grade level he would be teaching. Yet, he found himself hyper-focused on the candidate's resume and missed the opportunity to fully consider other skill sets essential to make the individual a successful teacher at the school he was part of.

- **The Horns Effect:** The Horns Effect occurs when individuals have negative aspects of their personality or behaviors you perceive as unfavorable, which alter the way you see them and overshadow the positive aspects they demonstrate.

For example, Pascal spoke to a supervisor of a large department in an organization. She was talking with him about an individual who was a leader on her staff. She pointed out that this individual frequently had typos in the messages he sent. Other leaders thought that the individual was an outstanding employee, but the supervisor focused mainly on the employee's negative aspects, which seemed to overshadow his positive attributes.

- **The Contrast Effect:** This effect happens when you compare an individual with an employee you hold in high esteem and miss out on the different strengths and talents the individual brings to an organization.

  This can easily happen when screening resumes, which is an arduous process. Imagine you have a candidate you really like and then use that candidate as the gold standard. You may miss different aspects of a person's experience that do not align with the gold standard, even though those attributes could be an asset to the organization. Essentially, you limit your basis of comparison to the few categories where your gold standard excels.

Understand that biases influence our decisions and may result in an unfavorable outcome. For example, we recall a time when we interviewed a candidate with our colleagues. The woman was African American and had a way of speaking that did not come across as what people at the organization deemed professional. However, she had her master's degree and qualifications to serve the position we were hiring for well. But, many on the panel were biased against her speech—the Horns Effect.

A day or so later, we all received thank you notes from her. Unfortunately, although the gesture was appropriate, we discovered she had used the same message for each of us, and in doing so, the same typo appeared on each note. These facts further solidified the judgment of her attributes because of her speech. Therefore, she was removed from our list of candidates.

Remedies for Organizational Bias

So, now that we understand how the brain works and the types of organizational bias, how can bias within an organization be eliminated or significantly reduced? The answer is simple and involves three basic steps:

1. The organization or some individuals within the organization may need a paradigm shift.

For example, author Tiffany Jana states that individuals must own their responsibilities and be a part of the solution. If people think they have no role in the overall environment they work in, they are sticking their heads in the sand. Every employee impacts the organization through words and actions and must be responsible for them. Any organization is a sum of its parts.[114]

2. The organization must have a systematic process to review data to eliminate organizational bias.

It is crucial to adopt diversity, have a good data culture, and have a process by which you analyze that data. Ellen Mandinach, a scholar and research scientist, sheds light on the best data collection and analysis practices. She calls her process a data-driven decision-making framework. Below are crucial points in her approach:

- Data in its raw form, without context, is meaningless.

- Data collection requires tools and an organized system that makes sense; one must know what data to gather and how to collect the data.

- Transform the data into information; data is put in context to provide meaning, especially as it pertains to different identity groups within an organization.

- Information gets transformed into knowledge about why certain effects are occurring. Determining the root cause will guide future actions.

- Based on a knowledge assessment, decisive action will be implemented and the impact assessed.[115]

3. The organization must actively seek to counteract bias using a framework or structured approach.

Individuals need to look for ways to counteract organizational biases using a structured process and framework. Authors Tiffany Jana and Ashley Diaz Mejias propose the "Erasing Institutional Bias

[114] Jana, T., & Diaz Mejias, A. (2018). *Erasing Institutional Bias: How to Create Systemic Change For Organizational Inclusion. First edition.* Berrett-Koehler Publishers.

[115] Mandinach, E. B. (2012). "A Perfect Time for Data Use: Using Data-Driven Decision Making to Inform Practice." *Educational Psychologist, 47*(2): 71-85. doi:10.1080/00461520.2012.667064.

Framework." This states that the journey towards being part of the solution to erase institutional bias requires courage. When individuals see bias within aspects of an organization, it will be hard to "unsee it." Their inaction may lead to guilt and shame that may cause them to slow down or give up on the work.

For leaders implementing DEI policies, here are some steps to consider when evaluating processes and procedures:

- **Set a clear intention and create a sense of urgency.** In other words, determine what is the goal of counteracting bias and the sense of urgency. Best-selling author, professor, and consultant Dr. John Kotter indicates that many companies fail to create a sense of urgency around their goals.[116] It may be important to clearly articulate the business and ethical case for counteracting bias and how that aligns with the overall mission and vision of the organization. Creating a sense of urgency involves examining the shortfalls that may occur if a goal or intention is not met. For example, consulting firm McKinsey demonstrated that organizations in the top twenty-five percent in racial and ethnic diversity were positioned to exceed their industry-specific median returns by thirty-five percent.[117] With these statistics in mind, it is crucial to recognize that organizational bias can result in a lack of diversity. Moreover, when there is a lack of diversity in an organization, attracting diversity can be a significant challenge because it may be challenging for underrepresented populations to feel a sense of belonging without people that share their identities.

- **Create a team that is able to analyze data.** It is important to gather and review facts about the organization. As we discussed before, the data you get has to be turned into information where the data is disaggregated by social categories, such as, but not limited to, race, ethnicity, gender, ability, etc. Having a diverse team, both in terms of identity, level, and departments, will enable people that have multiple perspectives to speak into the data, especially how it

[116] Kotter, John P. (1995). "Leading change: Why transformation efforts fail." Harvard Business Review 85(1): 96-103. Business Source Complete, EBSCOHost.

[117] Hunt, Vivian, Sara Prince, Sundiatu Dixon-Fyle, and Kevin Dolan. (2020). *Diversity Wins.* McKinsey.

pertains to groups of individuals and departments within the organization. This way, individuals feel as though they are not only raising issues within the organization, but they are part of the solution, which inevitably allows them to feel valued.

An organization Pascal recently consulted for formed a team dedicated to looking at and analyzing data pertaining to DEI. The team represented different parts of the organization, and they created a team charter, mission, vision for the group, goals for their workplace culture relating to DEI, and a scope of work. They collected data and considered how they were doing in relation to those goals.

- **Identify areas of growth and create a list of priorities.** The data may reveal areas of growth for the organization and allow the team to pinpoint reasons organizational goals were not being met. The information is typically communicated to an executive sponsor or key decision makers so they can prioritize the interventions they would want to implement. The beauty of collecting data using consistent qualitative or quantitative surveys is you can track progress and growth in key areas.

- **Communicate with stakeholders and be clear as to the role they play.** Once priority areas are identified, it may be important to create an executive summary document to communicate with stakeholders in the organization. Each individual has shared accountability to perpetuate behaviors that will achieve the desired culture they want to see.

In one particular organization, data revealed that groups of individuals did not feel connected, which is important for belonging. The DEI team felt as though people not feeling connected was a prerequisite for bias in the organization. So, the team challenged managers to come up with ways to facilitate connections within their teams. One leader created a weekly newsletter that highlighted personal accomplishments, anniversaries, birthdays, births, etc. This set the expectation that people should work to get to know each other not only professionally, but personally as well. It enabled people to slow down their thinking and judgments towards groups of people and consider who they were individually.[118]

- **Create structures to sustain change.** Just like the German company we previously talked about, they instituted a protocol where important decisions would include a devil's advocate who would ask tough questions about the proposed decision. They institutionalized this practice to allow for differences of opinion so that the organization could grow.

Many companies are doing blind resume screenings where they do not look at the name of the applicant but consider their credentials to institute more objectivity in the resume screening process. Moreover, many institutions are creating detailed and consistent practices, such as having a hiring committee, asking consistent questions during interviews, and asking for feedback shortly after the interview to reduce confirmation bias and groupthink. These structures allow sustained change from situation to situation.

- **Build accountability by continually checking progress.** Review the changes and ensure they are working as intended. Therefore, it is important to collect feedback and hold the organization and leaders accountable to their goals.[119]

For example, if mentorship and sponsorship are a priority for the organization, then leaders and stakeholders can be kept accountable by seeing whether they are participating in the mentorship and sponsorship programs and assessing whether the programs are effective. With that said, it is important to consider that failure to reach goals may indicate a shortcoming in structure and should not necessarily be put on the person or people immediately. Maybe communications were not clear, or there may have been turnover within a department. Accountability does not mean punitive measures should be pursued, rather that good questions are asked for the benefit of the program.

[118] Nelson, Olivia. (2017). "Potential for Progress: Implicit bias training's journey to making change."

[119] Jana, T., & Diaz Mejias, A. (2018). *Erasing Institutional Bias: How to Create Systemic Change For Organizational Inclusion: Vol. First edition.* Berrett-Koehler Publishers.

We recently read an example of how accountability can be used effectively. There was an organization that was looking to diversify their applicant pool. They changed their interview structure to include both in-person and take-home portions for the interview. They discovered that their applicant pool was not getting more diverse, but less diverse. The leaders paused the process to see what was going on so that the structure could be improved. As they looked more deeply into the issue, they discovered that they did not have a diverse talent pool before the position was posted, and they needed to spend more time building their talent pipeline in addition to making changes to the interview process.[120]

Individuals need to see themselves as impacting the organization they work for. Basically, employees must "buy-in" to their role in bias projection. Then an organization must make data the driver for change: research, information, knowledge, action. Of course, we cannot change what we do not acknowledge, so again, let data be the change agent. Finally, implementing solid framework processes will keep biases at bay, no matter the individual mix within an organization.

Organizational Bias, Interpreting Data, and Our Brains

CEOs by Race and Ethnicity

Asian/Indian   Black   Hispanic

Number of Fortune 500 and S&P 500 Companies with CEOs of this race/ethnicity group

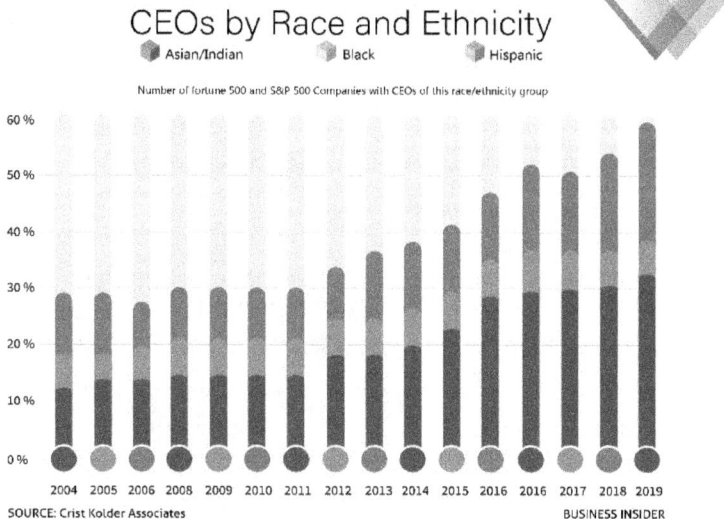

SOURCE: Crist Kolder Associates

BUSINESS INSIDER

---

[120] Carter, E. (2022). "DEI Initiatives Are Futile Without Accountability." *Harvard Business Review.*

107

How should we react to the previous chart? First, we must acknowledge the potential reactions some may have had to it. For example, one may look at the graph and think, "This is systemic bias and the result of the cumulative effect of individual bias." Or, they may think, "I see what the chart says, but there must be more to this story than just bias. We do not know the circumstances behind the numbers nor the details behind each hiring decision. Therefore, we need more context."

A 2018 study by Matthew Fisher and colleagues at Carnegie Mellon University revealed that we tend to have a binary bias when we look at data and information. In other words, we tend to reach conclusions that a particular effect is present or not, even in the presence of nuances. So, for example, when many people look at the data regarding representation in the business world, even though we may know there are nuances in the data, Matthew Fisher and colleagues suggest that we will tend to conclude the effect we see is a result of bias and that dominant narrative will become our reality, like we saw with the survey of CEOs and the definitive lack of minority individuals in the mix.[121]

Our minds will always attempt to classify and simplify things to make sense of them. Why? Because, again, our brains do not do well with uncertainty, ambiguity, and a lack of clarity. Those areas are rendered unsafe. Therefore, we are always trying to find certainty in uncertainty, and we are trying to find clarity where there is no clarity.

Following the first split-brain surgery in the 60s and 70s, a popular narrative emerged about the left and right hemispheres of the brain. It became popular to think that the left side of the brain-controlled logic, systematic thinking, and analysis, while the right-side modulated creativity, emotional intelligence, imagination, and intuition. You may have heard people say, "I am more left-brained dominant," meaning they are more analytical and logical. Others may have said, "I am more right-brain dominant," because of their creativity. Well, this line of thinking has since been proven false and is considered one of the many myths about the brain. In reality, both brain hemispheres work together very closely and communicate to solve problems when experiencing the world. However, although our left and right brains constantly work together, specific actions

---

[121] Fisher, M., & Keil, F. C. (2018). "The Binary Bias: A Systematic Distortion in the Integration of Information." *Psychological Science*, 29(11), 1846-1858.

produce more activity on the left side of the brain, and other measures make more activity in the right side of the brain.

The left side of the brain causes us to pay close attention to the things we find important and may lead us to not see the nuances in different people's arguments. So, when our left hemisphere is engaged, we tend to have a narrow view of the world.

Now, if you were to look at the brain's anatomy, the right side of your brain has a larger prefrontal cortex, so it can understand context, question things, and have different ways of thinking through things. The right side of our brains enables us to see individuals and individual perspectives and circumstances. When we engage the right hemisphere, we build connections and consider the world views of others.

Suppose we pull up the charts representing racial categories in the business world. In that case, the increased activity in your left hemisphere will pay attention to the charts' details. The right hemisphere will show increased activity as you extrapolate that data to your experiences and world views. Make sense?

As we look at generalizations and connect with the world, our brains have another mode that sets human beings apart from other species. Our large frontal lobe allows us to not only look at details and generalize but also step back from the data and consider other interpretations and perspectives that are different from our typical ways of thinking. For example, it allows us to empathize. Our frontal lobes exist to tone down our emotional brains and emotional reactions and reason through different situations.[122]

It may be important to consider that when we look at data, it is easy to interpret the data through our subjective lenses and make assumptions as to why particular effects are present. This section calls for us to do our homework, use our full brains, and not rely on our automatic frames of reference when looking at organizational data. As we learned earlier in the chapter, it is important to methodically look at data to pinpoint why certain trends are evident.

---

[122] McGilchrist, I. (2019). *The Master and His Emissary: The Divided Brain and the Making of the Western World.* Yale University Press.

Summary

When implementing systemic change in your organization, remember:

1) Change starts with your employees; they must be committed.

2) Develop systematic ways to engage with data:
    - Remember raw data without context is meaningless.
    - Create an organized system with precise tools for collection.
    - Transform data into information.
    - Transform information into knowledge, revealing root causes.
    - Choose actions based on knowledge.
    - Assess impact of choices.

3) Use a framework and structure:
    - Create a data analysis team.
    - Identify areas of growth and priorities.
    - Communicate with stakeholders about their role.
    - Establish structures to sustain the change.
    - Build accountability by systematically checking progress.

Chapter 5

⌘

Polarized Thinking

"Polarization causes distrust, and distrust causes polarization."
—Nolan McCarty

Kevin, Chief People Officer for Company A, heard a heated conversation between two groups of employees. The employees were discussing a new policy for inclusivity training, a new requirement for all employees or stakeholders in the company. One group of employees argued that inclusivity training only benefitted a small number of individuals and excluded the majority. The other group claimed the change in thinking or approach should have happened a long time ago. The intense discussion divided the groups into two factions—binary thinking. As more individuals joined in, they took the side they most aligned with. Think about this scenario and how Kevin should handle this situation as you read on.

Binary thinking means adopting a "my way versus your way" approach to dealing with an issue. For example, in America's current political and social climate, there are two schools of thought when discussing the issue of poverty. Some believe poverty results from an individual's choices, while others believe it results from circumstances beyond their control. In this case, binary thinking suggests that you are either for or against it, and there is no in-between. Unfortunately, we are a country divided, and each side has taken its stance, digging its heels into the sand and projecting its best arguments for or against it. This exemplifies how individuals in society and organizations can be polarized on issues, as we saw with the employees in the scenario from the previous paragraph.

When we think in binaries, those dichotomous categories are often in opposition. For example, supporters and opponents, allies or foes, good or bad, racist and anti-racist, Republican and Democrat. This places us in a "you are either with us or against us" society.[123] Research shows that binary thinking can lead to higher defensiveness, low humility, low honesty, and low agreeableness, which can affect an organization's productivity and effectiveness.[124] Unfortunately, the more divided a community, the more likely we will continue to think in binaries and be less willing to find common ground.

Binary Thinking in Organizations

Individuals bring their societal biases into any organization to which they belong. For example, our beliefs, values, and worldviews come with us when we work or attend school. But when we exist in a polarized and divisive context, it compromises trust among people in the organization. We may not be willing to compromise, find common ground, or dialogue about our differences in this environment. Instead, we may look for and point out the worst in each other and transform the organization into a "winner takes all" society. Basically, we pursue our self-interests and reinforce our polarized positions.[125]

In 2017, the United States government imposed a travel ban on Iran, Iraq, Libya, Somalia, Sudan, Syria and Yemen—predominantly Muslin countries—to suspend the entry of refugees from these countries. The ride-share company, Uber, faced tremendous backlash from its employees and customers because they did not speak up in protest of the ban. Many perceived their silence to mean they supported the ban and applied binary thinking to the situation.[126] In fact, reports showed that when taxi drivers at John F. Kennedy International airport in New York protested the

---

[123] Oshio, A. (2009). "Development and Validation of the Dichotomous Thinking Inventory." *Social Behavior and Personality: An International Journal*, 37: 729–741.

[124] Ashton, M. C., & Lee, K. (2007). "Empirical, Theoretical, and Practical Advantages of the HEXACO Model of Personality Structure." *Personality and Social Psychology Review*, 11: 150–166.

[125] McCarty, N., Poole, K. T., & Rosenthal, H. (2016). *Polarized America: The Dance of Ideology and Unequal Riches.* MIT Press.

[126] Reeves, M., Quinlan, L, Lefevre, M & Kell, G. (2021). "How Business Leaders Can Reduce Polarization." *Harvard Business Review*.

government's travel ban, Uber kept their services active and profited tremendously. The hashtag #DeleteUber was used to protest the organization, and the company released reports stating they lost hundreds of thousands of customers.[127] This, along with other scandals that followed, caused Uber's workplace culture to suffer.

In binary thinking, there is typically a dominant force. Scholars Kenneth Jones and Tema Okun discuss essential characteristics of the dominant culture. They explain that our society is marked by "either/or" thinking, essentially binary thinking. Learning from mistakes and working through conflicts is more difficult when we exhibit "either/or" thought processes because people are forced to take on and defend positions, rather than consider the nuances in people's arguments and perspectives.[128] Moreover, "either/or" thinking simplifies complex issues and can create a sense of urgency for organizations and their leaders to weigh in on political and social issues.

What can leaders do when faced with polarizing issues? We suggest the following:

- **Reinforce your values, especially those around dignity, respect, and civil dialogue.** An organization faced with polarization following the death of George Floyd issued the following statement to their community in order to reinforce its values around dignity, respect, and civil discourse:

    > Hateful, demeaning speech is fundamentally at odds with the type of school community we aspire to be. It makes others question their sense of belonging in our community of learning, makes it more difficult for them to engage meaningfully in community and conversation, and makes it impossible for us to sustain a culture of open, respectful discourse... As we navigate difference and disagreement, conflicts and misunderstanding will occur. Grace and healing are important in these moments...Grace is not a

---

[127] Leskin, P. (2019). "Uber Says the #DeleteUber Movement Led To 'Hundreds of Thousands' of People Quitting The App." https://www.businessinsider.com/uber-deleteuber-protest-hundreds-of-thousands-quit-app-2019-4.

[128] Okun, T., & Jones, K. (2000). "White Supremacy Culture." *Dismantling Racism: A Workbook for Social Change Groups,* Durham, NC: Change Work. Retrieved from http://www.dismantlingracism.org/Dismantling_Racism/liNKs_files/whitesupcul09.pdf.

moment of simple absolution, but a process of earning a deeper form of togetherness, one that makes us worthy of our highest ideals...[129]

- **Allow stakeholders within the organization to be part of the solution.** Oftentimes, when polarization happens it is important for leaders to speak with people in the organizations to not only gauge how they are feeling but also to brainstorm ways to move the organization forward in spite of apparent polarization.

Following the deaths of Philando Castile in Minnesota, Alton Sterling in Louisiana, and five police officers, Lorne Ahrens, Michael Krol, Michael Smith, Brent Thompson, and Patricio Zamarripa in Dallas in 2016, Tim Ryan, U.S. Chairman and Senior Partner at Price Waterhouse Coopers (PwC), received an email from a Black employee indicating that the firm had said nothing about the unrest that was happening following these deaths. Tim Ryan shut offices down for the day to have company-wide discussions about race with the goal of hearing from different stakeholders. The conversations were guided by norms for maintaining a respectful posture during these conversations, and the conversations proved to be very successful.[130]

- **Be proactive and not reactive.** An organization that waits for tragedy to foster intergroup contact may run the risk of being labeled performative, which means they are only engaging in dialogue and discussions over a particular issue to save face or limit criticism. It is important for organizations and their leaders to communicate and demonstrate their commitment to DEI on a consistent basis. Leaders and organizations can do this by having the following in place:

  - ○ **A DEI charter** that clearly indicates the goals of the organization and actions that could lead to the fulfillment of the goals.
  - ○ **Leaders who publicly and privately demonstrate their commitment to DEI.** This can be done through goal

---

[129] St. Albans School. (2022). *St. Albans' Vision and Expectations For Respectful School Discourse.* St. Alban's School. Accessed August 25, 2022.

[130] Gelles, D. (2021). "There Is A Bigger Role": A C.E.O. Pushes Diversity." *The New York Times.* https://www.nytimes.com/2021/03/05/business/tim-ryan-pwc-corner-office.html.

setting and organizational dashboards to illustrate progress on goals.

- ○ **Dedicated resources to advance DEI goals**. Often people look at financial resources as a measure of how committed an organization is to DEI.
- ○ **Structures to gather feedback from stakeholders**. This can include surveys, listening sessions, and DEI-specific questions on surveys.
- ○ **Employee Resource Groups.** These groups provide identity-specific spaces for individuals within the organizations to find support and contribute to the workplace.[131]

The Importance of "Why": the Purpose and Goals of the Organization

As people start to work towards a common goal, biases are reduced. Sixty-plus years ago, a classic study experiment called the Robbers Cave experiment was conducted. In the experiment, middle school boys of similar ethnic backgrounds and religious beliefs, who had never met each other, were randomly assigned to two groups. The two groups did not know of the other's existence and were separately taken to a camp called Robbers Cave. The groups were isolated from each other and were required to bond by performing different activities. As a result, each group developed strong loyalty and group ties.[132, 133]

Moreover, when the two groups met to participate in competitive activities, members of each group showed strong in-group bias. They characterized their group members favorably and the opposing group

---

[131] Jamal, N. & Tschida, T (2021). "3 Actions for Leaders to Improve DEI in the Workplace." *Gallup.* https://www.gallup.com/workplace/348266/actions-leaders-improve-dei-workplace.aspx.

[132] McLoed, S. (2008). "Robber's Cave Experiment." Simply Psychology. Retrieved July 30, 2019. https://www.simplypsychology.org/robbers-cave.htmlhttps://www.simplypsychology.org/robbers-cave.html.

[133] Sherif, M., Harvey, O. J., White, B. J., Hood, W. R., & Sherif, C. W. (1961). *Intergroup Conflict and Cooperation: The Robbers Cave Experiment (Vol. 10).* Norman, OK: University Book Exchange.

negatively. For example, researchers observed name-calling, derogatory songs, and even physical altercations.

The researchers tried to reduce the friction by combining the two groups and completing tasks as "one team." When the two groups were required to work together towards a superordinate goal, for example fixing the camp's water supply, the group started working better together. The experiment showed that social bonds form relatively quickly and demonstrated people's need to be connected. These social bonds form more easily when individuals have a common goal and vision that they can connect with. The Robbers Cave experiment shows that when there is a strong "why" or mission and vision that employees can anchor themselves to, they are more apt to work together towards the goal and cultivate connections to achieve a positive end result.

When individuals feel united, there will be a more significant movement toward cooperation over competition and a focus on positive interactions, problem-solving, and intergroup relations.

Yet, we must understand that when dialogue happens about opposing positions and the interests of different stakeholders, it may become apparent that other stakeholders have the same core interests but disagree on the mechanism or vehicle used to satisfy those interests. Therefore, it is essential to remind individuals that the ideal state may not be achieved. However, a better result can be achieved through intentional and pointed conversations.

Counteracting Bias Within the Organization

Considering what we have discussed in this chapter about binary thinking, we recognize that polarization has an impact on stakeholders within an organization and that many leaders have a sense of urgency to respond in the right way. As leaders create initiatives to meet the needs of stakeholders in these divisive times, it may be important for them to implement inclusive strategies that engage different stakeholders and thereby maintain a healthy workplace culture. We suggest that leaders:

- **Prepare for varying reactions to positions and initiatives.** Recognize that there will be different reactions to situations and initiatives, especially on diversity, equity, and inclusion. In order to mitigate the effects of uncertainty, it is essential for leaders to:

o Use data to communicate the current state of the organization and why the initiative will move the organization towards a desired future state.

o Communicate how the initiative connects to the overall mission and vision of the organization.

o Communicate the ways in which the initiative will impact the current state of stakeholders within the organization.

o Communicate the resources and supports that will be available to the different stakeholders for the new initiative.

o Show evidence-based research that supports the efficacy of the initiative or program.

o Present clear timelines or timeframes for the proposed initiatives.

o Communicate the metrics that will be used to demonstrate progress.

- **Assess the engagement and the level of influence of groups of individuals.** Every organization has stakeholders, and, within the stakeholder group, you have influencers, supporters, skeptics, and fence-sitters. Understanding how to engage these different groups will be instrumental to the success of your organization, especially when implementing new initiatives. We will discuss each one in detail in the following sections.

Stakeholders

Stakeholders are the various groups and individuals who will influence the success of your new plans. These individuals are invested in the organization where the change is occurring. In addition, it is often the skills used to communicate, consult, and involve these people which will determine an initiative's success. Therefore, it is essential to know how to manage different stakeholders. First, note that there will be varying degrees of engagement and interest within these groups of people that may change during your initiatives. To understand what we mean, we will list the characteristics of the two levels of engagement found in stakeholders: high and low.

First, we would like to address high-engagement stakeholders. We have had the privilege of building Diversity, Equity, Inclusion, and Belonging initiatives from the ground up. As we structured these initiatives, there were highly-engaged individuals who had a sense of urgency for the work. When we reflect on why these individuals had a sense of urgency, their experiences caused them to connect on a deeper level and empathize with the work of inclusion and belonging. For example, a few individuals had terrible school experiences and expressed deficit narratives of themselves and their identities. Their feelings of unbelonging compromised their ability to focus, engage, and inevitably reach their full potential. Other individuals had loved ones, particularly children, affected by cultures and environments of unbelonging.

For the highly-engaged individuals, the work of belonging was not a fad but one that affected their well-being and abilities to reach their personal and professional goals. Here are some common characteristics we observed from highly engaged individuals:

- These individuals possess a personal connection to the initiative. More specifically, they are personally affected by the issue(s) the proposal or measure is addressing.

- They know someone that has been affected by the issue(s) being addressed.

- They want the organization to be unified and productive.

- They possess a passion for learning, are intrigued by the initiative, and want to be part of it in any way possible.

Although being around the highly-engaged stakeholders was empowering for us, other individuals were not as committed. We refer to these people as low-engagement stakeholders. We distinctly remember a few colleagues talking to us about a recently-attended training. We gathered that the leaders who conducted the training did not understand our industry and lost credibility with the participants. Many of our colleagues left asking what the "hidden agenda" was and distrusted the intentions of the organization's leadership. Other individuals had what we like to call the "here we go again" mindset. These individuals had been part of similar initiatives, and nothing seemingly came from the lessons and conversations they learned. As a result, they were not as engaged because the prospect of being disappointed and deflated outweighed the possibilities of progress and change.

Here are some characteristics of the low-engagement stakeholder:

- They are not invested in the organization, may have other priorities, or just want to do their job and leave.
- They have seen many initiatives come and go with no benefit to them.
- They are disconnected from the work and have no personal stake in it.
- They are concerned about how the initiative will affect their roles and responsibilities.
- They do not trust in the organization.
- They have a fear of the unknown.

Influencers

Every organization has stakeholders deemed influencers, who are individuals having a significant impact on the organizational environment. For example, in one of the organizations we were leading, a gentleman had been with the company for many decades. He identified as a conservative and had a heart for engaging and motivating people. He was someone that many people admired and looked up to. Being able to "speak his language" was vital for us to get buy-in from him, and we knew that if we had his support, others would follow.

It is important to note that even though the gentleman we referenced was a clear influencer, it is essential to do your homework so you know who these people are because influencers may not always be the loudest people in the room. Furthermore, they are not always who you think they are. Typically, influencers administer their opinions through the grapevine. Often, these conversations are spearheaded by a small but mighty few. According to recent research, three percent of employees, "the mighty few," affect ninety percent of the essential conversations and discussions that occur in an organization.[134] These are individuals who have won trust and respect in an organization. As a result, employees seek out and value their advice as they provide information about what is happening in the company.

Some best practices in utilizing these individual's influence for the good of the organization are:

- Bring employees with influence into your initiative because they can be champions for your initiative and affect change in ways

---

[134] Duan, L., Sheeren, E., & Weiss, L. M. (2014). *Tapping the Power of Hidden Influencers.*

other stakeholders cannot. In essence, they can create more buy-in.

- Encourage influencers to translate any changes into relatable terms for colleagues in the organization, something they are well-suited to do because they often relate better to other employees than leadership can.

- Invite influencers, especially those who buy into an initiative or change, to provide valuable insights and feedback through formal and informal dialogues to gauge how people in the organization feel.[135]

However, not all influencers are the same. Below, we have detailed the different types of influencers.

First, the type one influencers are high engagement and high influence. We recommend making a list of individuals you believe are influencers with high engagement and high influence, and then follow the steps below to work toward unity:

- **Recruit:** Put out communications that you are looking for people passionate about an initiative and *recruit* them to play an active role.

- **Recommend:** Identify passionate influencers and ask them to give you a list of more people they recommend, creating a snowball effect.

- **Root:** Communicate to bring these influencers into the conversation and root them into the initiative; ask them to help process feedback, assist with communications, or provide feedback on implementation plans.

- **Rally:** Unleash influencers to advocate for the initiative in their spheres of influence and rally interest.[136]

Next, you will look for the type two influencers, low engagement and high influence. First, make a list of individuals you believe are in this category, and then follow the steps below:

[135] Stople, M. (n.d.). *3 Ways to Identify Employees with Influence at Work.* Bonfyre. February 2, 2022. https://bonfyreapp.com/blog/identify-employees-influence-at-work.
[136] The alliteration "Recruit, Recommend, Root, and Rally" codeveloped with Carolyn Ovitt.

- **Relate:** Intentionally meet with these individuals to form relationships, learn from them, and communicate expectations.

- **Recommend:** Identify these influencers who are not passionate. Ask them if there are any more people they recommend you talk to.

- **Rotate:** Frequently communicate with them and allow them to rotate into some aspects of decision-making.

- **Reinforce:** Be intentional about demonstrating the initiative's impact to reinforce its importance.

Supporters

The second type of stakeholder is supporters. These include the people who frequently say, "Put me wherever." These individuals are happy to be part of the initiative and contribute in small ways. Supporters are not as persuasive as influencers, but they function in an organizational environment as "reinforcers" and are willing to act on a vision. Below is the type of supporters you must identify and how to best utilize their attributes.

There are supporters who demonstrate low influence with high interest. This group of people may not have the most significant impact, but they are supportive. Make a list of individuals that you believe will back the initiative, and then:

- **Recruit:** Put out communications that you are looking for people to support the initiative; recruit them to back it.

- **Recommend:** Identify passionate supporters and ask them to give you a list of more people they recommend.

- **Relay:** Keep them in the loop regarding initiatives and regularly relay updates to support the work and bring others along.

- **Routine:** Allow them to be involved in regular ways with the initiative.[137]

---

[137] The alliteration "Relate, Recommend, Rotate, Reinforce, Recruit, Recommend, Relay, and Routine" codeveloped with Carolyn Ovitt.

Skeptics

Next are the skeptics because they exhibit low interest and engagement. Skeptics are not engaged in the initiative and express low interest, but they can provide you with meaningful data to strengthen the initiative. List those potential skeptics in your organization, and then:

- **Respond:** Anonymous feedback can be a vital response mechanism for these individuals to voice their concerns. Open Q&A forums are also an excellent way to share their voices without judgment.

- **Report:** Frequently communicate with them and report back about how their feedback was considered.

- **Repeat:** Keep this group informed through repetition of the "whys" throughout the initiative.

Fence-Sitters

The other group is the fence-sitters. These individuals demonstrate low engagement with moderate interest. Try and do things to increase their interest in the project, such as:

- **Relate:** Humanize the initiative with anecdotes from people that have gone through issues similar to what the initiative is addressing, and help them *relate*.

- **Reach out:** Personally invite them to activities or meetings for the initiative—offer a relationship. Even if they do not attend, they may appreciate the gesture.

- **Ruminate:** Frequently communicate with them and offer opportunities to ruminate and process initiative concepts.

- **Repeat:** Keep this group informed through repetition of the "whys" throughout the initiative.[138, 139, 140, 141]

We All Need to Belong: Targeted Universalism

You may have read the above paragraphs and thought to yourself: *Why engage skeptical people who lack support for initiatives?* This is an excellent question. The simple answer is when it comes to race and ethnicity, "belonging" impacts us all. Whether you are an influencer, supporter, skeptic, or fence-sitter, we all need to feel included in a community.

This approach of including all stakeholders can be referred to as targeted universalism. Targeted universalism asserts that there must be a universal goal, the "universalism" part. The "goal" here is engagement and belonging for stakeholders in an organization. This approach's targeted portion means that different groups should not be ostracized but brought into the fold with precise strategies. In essence, leadership must understand and address the needs of all groups.

Targeted universalism can be very effective because the strategy minimizes both passive and active forces that result in exclusion or marginalization, especially of the majority group, in favor of all groups achieving the objective of engaging and belonging. So, in addition to using the strategies outlined above for the different stakeholder groups, it is critical for leadership to implement the five steps of targeted universalism:

1. Be clear about what the goal is. In this case, the goal is belonging.
2. Assess how the different stakeholder groups are doing in relation to the goal.
3. If certain groups are more engaged or have more of a sense of belonging compared to other groups, assess why.

---

[138] University of Virginia. (n.d.). *Organizational Excellence: Partnering for Effectiveness and Excellence.* February 1, 2021. https://organizationalexcellence.virginia.edu/change-management.

[139] Fisher, R., Ury, W. L., & Patton, B. (2011). *Getting to Yes: Negotiating Agreement Without Giving In.* Penguin.

[140] Stople, M. (n.d.). *3 Ways to Identify Employees with Influence at Work.* Bonfyre. February 2, 2022. https://bonfyreapp.com/blog/identify-employees-influence-at-work.

[141] The alliteration "Respond, Report, Relate, Reach out, Ruminate, and Repeat" codeveloped with Carolyn Ovitt.

4. Look at ways to better support each group.
5. Implement strategies that will enable the groups to inch forward toward the goal.[142]

Our Shared Humanity

In the current climate of polarization, there have been intense debates and divisiveness over Critical Race Theory (CRT) in schools. CRT is an academic framework that was introduced by legal scholars in the late 1970s and early 1980s and is considered part of what scholars called postmodernist thought.[143] Postmodernism is a movement that emerged to question universal truths and emphasizes that the world is subjective and reliant on the perspective of individuals. Therefore, those in power can create dominant societal narratives, even at the expense of those not in power, and those narratives often persist as universal truths. To illustrate this point, let us consider the 1619 Project. There has been huge criticism of the 1619 Project, a body of work dedicated to "reframe the country's history by placing the consequences of slavery and the contributions of Black Americans at the very center of our national narrative."[144] Many consider this work to have postmodern roots because it challenges the predominant narrative of America being the "land of the free and home of the brave." The project is also said to be an offshoot of critical theory, a Marxist-inspired movement, which studies systems of oppression and domination in order to upend the social structures that lead to the oppression of different groups.[145]

CRT, as a concept, emphasizes the fact that race is a social construct and asserts that racism does not exist only at the individual level, but is institutionalized and embedded into systems, such as legal, education, housing, medicine, business, etc. These systems perpetuate biased laws and policies that stifle the progress and upward mobility of people of color. For

---

[142] Powell, John, Stephen Menendian and Wendy Ake. (2019). "Targeted Universalism: Policy & Practice." Haas Institute for a Fair and Inclusive Society, University of California, Berkeley. https://www.haasinstitute.berkeley.edu/tar-geteduniversalism.

[143] Sawchuk, Stephen. "What is critical race theory, and why is it under attack?" *Education Week* 18 (2021).

[144] Hannah-Jones, Nikole, Mary Elliott, Jazmine Hughes, and Jake Silverstein. (2019). *The 1619 Project: New York Times Magazine.* August 18, 2019. New York N.Y: New York Times.

[145] Britannica, T. Editors of Encyclopedia. "Critical theory." *Encyclopedia Britannica.* https://www.britannica.com/topic/critical-theory.

example, many point to laws and policies in the K-12 education system that have led to what many term the resegregation of schools, underfunding of majority Black and Latinx school districts, inequitable discipline practices, and unbalanced special education referrals.

Over the past several years, topics like housing segregation, biased criminal justice laws, and the enslavement of Black people have entered the consciousness of society and have been emphasized by many school curriculums. Many critics of CRT state that its foundational tenets are divisive and unjustly labels Whites as inherently racist. Opponents of CRT also suggest that an organization's commitment to diversity, equity, and inclusion is a zero-sum game where people of color benefit at the expense of Whites.[146] A recent survey by the group Parents Defending Education indicated that some schools were teaching students that "White people are inherently privileged, while Black and other people of color are inherently oppressed and victimized." They also indicated that schools were teaching that "achieving racial justice and equality between racial groups requires discrimination against people based on their whiteness." Furthermore, they indicated that schools were teaching that "the United States was founded on racism." Their fear was that White students would feel shame and guilt and have their self-esteem and self-concept suffer due to the foundational concepts of CRT.[147]

In a Supreme Court Case in 2007, parents in the Seattle school district protested using race as a category to allocate students to certain schools in the district. The Supreme Court ruled 5 to 4 that relying solely on race to make school assignments, in order to diversify them, especially in tie-breaker situations, violated the equal-protection clause of the 14th Amendment. As part of the case, Chief Justice Roberts explained: "The way to stop discrimination on the basis of race is to stop discriminating on the basis of race," to which the late Justice Ruth Bader Ginsburg replied, "It's very hard for me to see how you can have a racial objective but a nonracial means to get there."[148] In other words, Justice Roberts is saying that we have to stop

---

[146] Jacob, Kathryn, Sue, Unerman and Mark, Edwards. (2020). *Belonging: The Key to Transforming and Maintaining Diversity, Inclusion and Equality at Work*. London; New York: Bloomsbury Business.

[147] Parents Defending Education. (2021). *Parents Defending Education National Poll: Americans Overwhelmingly Reject "Woke" Race and Gender Policies in K-12 Education*. August 24, 2022.

[148] Sawchuk, Stephen. (2021). "What is critical race theory, and why is it under attack?" *Education Week* 18.

emphasizing race to eliminate racism, whereas the late Justice Ginsburg is suggesting that we have to address the evils of racism and the legacy of inequality for people of color. Moreover, she is suggesting that in order to heal and move forward, we must look at context and work to restore equality through equity. That means for organizations and systems to direct resources towards elevating historically marginalized populations due to the troubling history of our nation.

Many supporters of diversity, equity, and inclusion work assert that it is important for children to be taught the truth about our country's past so they are empowered to change the future. Many educational leaders, legislators, students, parents, and guardians are advocating for moving from the traditional Eurocentric lens of history, science, and other aspects of the general curriculum to one where the whole truth of the past is taught and where the contributions of non-Whites are brought to the forefront. Many assert that this lens will enable students of color to make sense of society and be affirmed by the grit, perseverance, innovation, and creativity of their ancestors and question, rather than always celebrate, the complex history of our nation.

As you read the debate over CRT, you may have feelings for and against the arguments that were presented above. In the midst of your emotions and feelings, we cannot forget our shared humanity. Our shared humanity begins with the understanding that there is more to a situation than meets the eye. Every behavior is motivated by a desire to fulfill specific human needs. As we explained earlier, Maslow describes our primary needs include physiological needs, psychological needs (safety, belonging, and esteem), and purpose needs (self-actualization and self-transcendence). Even though these needs are universal, individuals have different experiences, perspectives, contexts, and truths that motivate them to fulfill these universal human needs differently. When we begin to view ourselves and others from the lens of our shared humanity, we may feel compelled to practice self-awareness through reflection and introspection and begin to understand that we, in our humanity, share similar needs, desires, impulses, thoughts, and emotions. Consequently, we must hold ourselves accountable for our thoughts and actions and treat each other with dignity. Our shared humanity requires humility and an acknowledgment of the uniqueness of others—this calls for us not to lead with assumptions and preconceived notions of who we believe people to be.

It is easy to make assumptions based on the beliefs that people hold on the topic of CRT. Maybe it is not CRT, but another divisive topic. What we can do is understand each other's background and experiences through respectful dialogue because the experiences, perspectives, beliefs, and values we hold are not easily observable on the surface but are essential components of who we are. Thus, we can perspective-take and develop a deep understanding of those around us, even if we may not share or fully understand where they are coming from. We will also understand that every individual desire leads back to the universal needs we all share.

Through the perspective of our shared humanity, we understand the need and the power of the human connection, thus the importance of working interdependently towards unity and reconciliation while breaking down barriers that stifle progress. Shared humanity compels a call to empathy and compassion with action—that is, to listen well to each other so we can appreciate their perspectives and learn to support one another.

While you might criticize us for living in a utopian world, we would like you to consider the fact that the needs parents and guardians are communicating when it comes to CRT are psychological safety, belonging, and (self) esteem for their children. Therefore, the debate over CRT is not only focused on what is being taught, but how it is taught. If the goal is for all parties to feel as though they belong, then the principles of Targeted Universalism suggest that different populations need different things. If affirmation and a positive self-concept is the goal, then it is important to consider what it will take for each group to reach that goal despite what is being taught. When we focus on functioning at the highest level of our human potential—with love, empathy and compassion—we can achieve unity and continue to move forward towards racial healing.

On a practical note, what should you do as a leader and educator if somebody asks whether you are teaching CRT in your school or classroom?

- CRT has taken on many varied definitions, so it is important to make sure you clarify what they mean by CRT. We have tried to be as objective as possible in our descriptions above.
- It is advisable to not be defensive, defining what you are not teaching and describing what you are teaching and the philosophy behind why you teach it, especially as it pertains to the school's values.

- Fear and other emotions are focused, so it is important to expand their thinking by comparing and contrasting other topics of study. For example, a school wrote the following in response to CRT criticism: "…While students learn about concepts related to CRT as one among many topics in current political and social discourse, they are not being taught CRT any more than they are taught fascism or socialism when studying European history…"[149]

Summary

When some people in an organization support DEI changes and others do not, try the following:
1) Reinforce the group's values and expectations for members.
2) Understand the thoughts and beliefs driving each side.
3) Communicate with all the "why" behind the initiative.
4) Determine who the influencers and supporters are.
5) Assess engagement of individuals and subgroups.
6) Allow stakeholders to be part of the solution.
7) Be proactive and not reactive about DEI.
8) Keep shared humanity in mind; listen for the needs people are communicating through their opinions.

---

[149] Columbus Academy (community message, July 12, 2021).

Chapter 6

Subtle Communication

"Words will scratch more hearts than swords."
—Atticus

O n a recent walk during a break at work, a younger man coming from the other direction happened to be wearing a t-shirt from Pascal's alma mater. Pascal stopped and asked, "Hey, did you go there as well?" The man immediately smiled and answered with an emphatic "Yes!" Over the next few minutes, the two discovered that they both attended the university for their undergraduate studies. At one point, Pascal tried to recall the name of his freshman dorm but could not. The younger man looked at him and said as he laughed, "Don't worry, we all have 'senior moments.'" Pascal returned the laughter as he understood the other man meant no ill-will, but he could not shake his comment the entire day. It stuck with him.

So, what are those subtle, seemingly innocuous little jabs, or slights if you will, that one person gives to the other, either intentionally or not? They are called microaggressions.

The Brain's Role in Microaggressions

The brain is an anticipatory organ constantly filtering information and assessing the data based on prior experiences. For example, if a man holds the door for a woman behind him, one woman may appreciate the gesture while another may be offended. How each individual analyzes a situation in their brain goes back to their lived experiences. So, although microaggressions are subjective, they are more about how the words or actions are perceived, meaning how they are processed in the brain.[150]

When individuals experience something, a signal goes to the part of their brain called the thalamus. The thalamus then distributes that signal to the amygdala, the emotional center, before heading onto the prefrontal cortex, the higher-order thinking processor. There is a much shorter distance between the thalamus and the amygdala. This explains why humans tend to be emotional beings—feel first, then think.

Once the amygdala receives a signal, it looks to the hippocampus, where the short-term memory resides, and asks, "Have I seen this before?" A stress response is mediated by the brain's HPA (Hypothalamus-Pituitary-Adrenal) axis if the answer is "yes." The HPA axis emits the stress hormone cortisol to "prime for action." This is the fight or flight response being ignited.[151]

Short or quick cortisol releases are favorable for an individual's safety—a human's internal security system. But when the cortisol "hangs around" too long, it can affect memory, the immune system, and your ability to regulate stress. Cortisol shuts down the higher-order thinking part of the brain, the prefrontal cortex, and keeps individuals in their emotional brain. As a result, people may become defensive or even avoid conflict altogether. Research shows the effects of cortisol—the reactivity, sensitivity, and negativity—can last for several hours.[152] The more one dwells and meditates on a situation, the more cortisol is released. As a result, cortisol can reduce an individual's ability to connect, show empathy, or provide the benefit of the doubt.

For example, a study was done where researchers surveyed 3,570 white-collar professionals. The survey showed that seventy-eight percent of Black people were fearful of being discriminated against at work or feared a loved one would be discriminated against. This percentage was three times more than White professionals. Why? It all goes back to experiences and how the brain interprets information.[153] People of color have witnessed or heard of

[150] Atzil, S., Gao, W., Fradkin, I., & Barrett, L. F. (2018). "Growing a Social Brain." *Nature Human Behaviour*, 2(9): 624-636.

[151] Kim, E. J., Pellman, B., & Kim, J. J. (2015). "Stress Effects on the Hippocampus: A Critical Review." *Learning & Memory*, 22(9): 411-416.

[152] Hannibal, Kara E., and Mark D. Bishop. (2014). "Chronic stress, cortisol dysfunction, and pain: a psychoneuroendocrine rationale for stress management in pain rehabilitation." *Physical Therapy* 94(12): 1816-1825.

[153] Hewlett, S., Marshall, M., & Bourgeois, T. (2017). "People suffer at work when they can't discuss the racial bias they face outside of it." Harvard Business Review.

others in their racial group having negative experiences. Those memories stored in their hippocampus and amygdala are accessed when the external signals are received by the "emotional brain." Since the hippocampus is rich in cortisol receptors, people will likely remember stressful situations with strong emotions.

The hippocampus also plays an instrumental role in shutting down the stress response modulated by the HPA axis—it calms an individual's response down. The hippocampus slows down the HPA axis by reducing cortisol production. However, when the hippocampus is damaged, it cannot do its job. Chronic stress, for example, can damage the hippocampus. Excessive cortisol production can cause the hippocampus to shrink. This impacts memory and results in less inhibition of cortisol production, which causes more hippocampal shrinkage—a vicious cycle.[154] In the end, people exposed to high-stress levels have their brains rewired to leave them with fewer cognitive resources, negatively impacting their ability to think, focus, be creative, and engage.[155]

Types of Microaggressions

Microaggressions are "brief and commonplace daily verbal, behavioral, and environmental indignities, whether intentional or unintentional, that communicate hostile, derogatory, or racial, gender, sexual orientation, and religious slights and insults to the target person or group."[156] That is a long-winded definition, so to clarify, let us break these aggressions down into three specific types: microinsults, microinvalidations, and microassaults:

- **Microinsults:** communication considered rude or insensitive. A microinsult demeans a person's identity and articulates that a particular group is not respected. Microinsults are often not intentional but can communicate subtle messages linked to unconscious biases about a person or group.

---

[154] McEwen, B. S., & Lasley, E. N. (2002). "The end of stress as we know it." Joseph Henry Press.

[155] Bremner, J. D. (1999). "Does stress damage the brain?" *Biological Psychiatry, 45*(7): 797-805.

[156] Sue, D. W., Capodilupo, C. M., Torino, G. C., Bucceri, J. M., Holder, A., Nadal, K. L., & Esquilin, M. (2007). "Racial microaggressions in everyday life: Implications for clinical practice." *American Psychologist, 62*(4): 271.

For example, if someone says to a person of color, "I am surprised you got that job since you did not meet all of the qualifications," the subtle messaging is that first, you are not qualified for the job, and second, you probably got the position for other reasons, such as race or ethnicity.

- **Microinvalidations:** communications that undermine, dismiss, ignore, exclude, negate, or nullify people's thoughts, perspectives, and experiences, particularly underrepresented groups.

For example, when someone tells another person, "I don't see color," it may be perceived that the individual may consciously or unconsciously negate or undermine the other person's racial experiences and cultural heritage.

- **Microassaults:** verbal or non-verbal attacks that overtly aim to hurt the recipient. The factor that distinguishes microassaults from microinvalidations and microinsults is the intentional nature of the act.

We watched a documentary where a mixed-race woman was recounting a situation where she was a server at a restaurant. While working one day, an individual went to the kitchen and told the woman that she would be in charge of serving the table of raccoons that had just walked in. In looking at the table, the woman discovered a table of Black individuals. This was an overt act of discrimination to compare Black people to animals, particularly raccoons, considered scavengers.

In another situation, an African American man indicated that during his days at the university, an acquaintance came up to him and said, "You are the least scary Black man I have met." This statement was considered problematic because there was an assumption of being scary and violent consistent with stereotypes that have been given to Black individuals, especially men, throughout history.

Microassaults include using symbols of hate, like a swastika, and other racial epitaphs that communicate violent, discriminatory language.[157]

Intention vs. Impact

A 2019 survey was given by *Fortune*, where researchers asked 4,275 individuals about their experiences with microaggressions.

- Forty-eight percent of those surveyed had experienced or potentially experienced microaggressions.
- Sixty percent of those surveyed had witnessed or potentially witnessed a microaggression.
- Ten percent of those surveyed believed they had committed a microaggression (which may, in part, indicate the acts were unintentional).[158]

Take a minute. Think about something you might have said to a colleague and afterward, you wondered if it could have been taken as offensive, despite your intention to compliment that individual. For example, suppose you, as a female, said to another female co-worker, "Wow, that dress makes you look thin." The co-worker immediately comes back with an awkward smile and forced laugh before uttering, "Thanks." However, after she walks away, you realize she may have interpreted your words as "she is not thin" or that "she usually does not appear thin." Either way, you think you may have insulted her unintentionally.

Now, place yourself on the receiving end this time. Imagine a colleague attempting to congratulate you on a job well done. The person walks up to you smiling and says, "You're not really Black. You're committed, consistent, and very articulate." The implied meaning might be received as Black people lack commitment and perseverance and are inarticulate—compliment not taken.

So regardless of intention, in either example, the impact may have been detrimental and lasting depending on the recipient's interpretation. This

---

[157] PBS Learning Media. (n.d.). "Microassaults, Microinsults and Microinvalidations." https://wosu.pbslearningmedia.org/resource/cb19-ss-types.microaggressions/microassult-microinsults-and-microinvalidation/.

[158] Gebhardt, J. (2019). "Microaggressions in the workplace." SurveyMonkey: Curiosity at Work. https://www.surveymonkey.com/curiosity/microaggressions-research/.

brings about the idea of intention versus impact: what one person meant versus how the recipient perceived the words.

This introduces an interesting concept: does the statement hurt less if the person saying it did not intend to offend? No. For example, if you fall and get a cut, the cut hurts. Now, will that cut hurt less if you fell by accident or were tripped by someone? In other words, do the specifics of how you fell make the pain any less? No. The same is true with microaggressions.

Therefore, a conflict between intention and impact can occur when the person delivering the words has a different idea than the person receiving them. This inequity of interpretation can lead to Fundamental Attribution Error, or FAE. FAE is when we attribute the slight or insult to a person's character and personality rather than the situation.[159] The thing about intention is that in the absence of listening and relationship, we can be subject to assuming intention based on how the other individual made us feel.

For example, when the "giver" commits a microaggression, the "receiver" may feel hurt and angry. In this instance, the receiver may link the aggression to the giver's nature, character, or identity, whereas the giver may view the receiver as too sensitive or overly critical.

Let us go back to the "compliment" given on a job-well-done previously to further explain. The person says to you, "You're not really Black. You're committed, consistent, and very articulate." You are offended and may assume this person has a low opinion of Black people and is a racist. Conversely, if the individual who made the statement sees it was taken as an offense, they may assume, based on their intent, that you are too sensitive.

When fundamental attribution errors occur, they result in a confirmation bias, where our reference points about types of individuals get validated, and no common ground is found. If we choose to close off communication and understanding, nobody grows to appreciate the other side's experiences, stories, and perspectives. There is no opportunity for a reinterpretation or reconciliation of the event. This situation leaves divisiveness to manifest and grow.

[159] Simmons, M. (2018). "Fundamental Attribution Error: This Cognitive Bias Destroys Relationships." Accelerated Intelligence. https://medium.com/accelerated-intelligence/fundamental-attribution-error-this-cognitive-bias-destroys-relationships-6f405895b81b.

Battle for Power and Moral Credentialing

When individuals feel accused, hurt, or have a sense of injustice, they often try to seize power in those interactions.[160] Individuals tend to hold biases towards their positive self-image, meaning they see themselves one way and feel others must view them in the same light.[161] That is part of why people love surrounding themselves with others who make them feel good. Because of this fact, when individuals are accused of dispensing a microaggression, they may react by:

- Attempting to seize power in interactions by justifying their intentions and becoming defensive.
- Liberating themselves from a label by citing examples of nonprejudicial behavior, stating things like, "My best friend was Black growing up," or "I have many sisters, and I am very close to my mother."

While someone is entitled to justifying their actions or "moral credentialing,"[162] the conversation cannot stop there. A few consequences may be realized if the conversation stops with one party defending themselves. These adverse effects are classified into the following categories: testimonial injustice, hermeneutical injustice, and contributory injustice.

- **Testimonial Injustice:** There are two types of testimonial injustice: testimonial quieting and testimonial smothering.
  - **Testimonial quieting:** This occurs when the person that performs an injustice refuses to hear or acknowledge the experiences of the individual impacted by the microaggression.
  - For example, one student says to another with a laugh, "You're Asian. Of course, you got an 'A.'" The Asian student replies, "That's racist and stereotyping. It hurt my feelings. I work hard

---

[160] Hardy, K. (2018). "DCF Racial Justice Summit 2018 Keynote." YouTube. https://www.youtube.com/watch?v=lRRQR6ZGCPc.

[161] Fields, E. C., Weber, K., Stillerman, B., Delaney-Busch, N., & Kuperberg, G. R. (2019). "Functional MRI reveals evidence of a self-positivity bias in the medial prefrontal cortex during the comprehension of social vignettes." *Social Cognitive and Affective Neuroscience, 14*(6): 613-621.

[162] Brown, R. P., Tamborski, M., Wang, X., Barnes, C. D., Mumford, M. D., Connelly, S., & Devenport, L. D. (2011). "Moral credentialing and the rationalization of misconduct." *Ethics & Behavior, 21*(1): 1-12.

for my grades." The person who committed the microaggression laughs it off and walks away, showing they did not hear the other—no validation.

- o **Testimonial smothering:** This happens when the individual receiving the slight tells a partial truth about how they are impacted by the microaggression. This may occur because the recipient anticipates the person who imparted the comment will not fully acknowledge and accept their testimony.

  Using the same example above, the Asian student may reply, "I have to work for my grades, too." In this case, the receiver of the aggression withheld their true feelings to ward off a defensive reaction.

- **Hermeneutical Injustice:** This refers to an inherent lack of diverse voices in leadership or positions of influence. This may allow for characterizations of people and behaviors from other cultures. If leaders do not interpret or see their characterizations as biased, their beliefs will not be addressed. The recipients of these biases may not have a venue to articulate wrongful social experiences. The result is a misunderstanding and misinterpretation, and ignorance may be preserved.

  For example, suppose a young Latino man enters an all-White school. The leaders of that institution may not feel there are any biases, but then suppose a White student says to the Latino student, "How did you end up coming to this school?" The Latino student feels the impact of the micro-aggressive statement, but he says nothing because he has no outlet to express his emotions.

- **Contributory Injustice:** This is when individuals in an organization dare to come forward and share their experiences. Nothing systemic or at the level of the organization is done to take those testimonies, understand them, and make the necessary changes within that organization. The hallmark of contributory justice is the acknowledgment of issues without action taken. As a result, the known problems remain unresolved for various reasons, like discomfort, expensive cultural shifts, and time investment.[163]

The consequence of testimonial injustice, hermeneutical injustice, and contributory injustice is "battle fatigue." When individuals undergo microaggressions, they may feel like they do not belong. As a result, they may exhibit low motivation, engagement, and productivity. As we have learned previously, unbelonging can lead to behavioral issues, higher attrition rates, and increased stress, depression, and anxiety. In fact, these individuals are more vigilant about monitoring their environment for signs of unbelonging (a high-context-dependent state) than focusing on the work.[164]

Additionally, individuals within an organization often look to their superiors to gain a positive sense of self and use their superiors' perception of them to assume a value or worth. When individuals feel like their leaders do not have a positive perception, or if they cannot change the presumed perception of their environment, battle fatigue may be experienced. The employee can get burnt out and eventually quit or leave the environment for good.

In a survey on microaggression, sixty-seven percent of the participants indicated that they preferred the person who committed the microaggression to recognize it and apologize. Of those who experienced that situation, eighty-three percent said that the apology was accepted. At the end of the day, it does not matter whether the microaggression is intentional or not. The important thing is for the person communicating the aggression to see and acknowledge the impact of their words.[165]

Social Cultures and Microaggressions

In society, there are two cultural ways of being that people can lean toward: individualism and collectivism. If you were to think of it like a continuum, individuals typically lean toward one or the other, which can impact perceptions of what is and is not a microaggression.

---

[163] Applebaum, B. (2018). "Listen! Microaggressions, Epistemic Injustice, and Whose Minds are Being Coddled?" Philosophy of Education Archive, 2018(1): 190-202.

[164] Smith, W. A., Hung, M., & Franklin, J. D. (2011). "Racial battle fatigue and the miseducation of Black men: Racial microaggressions, societal problems, and environmental stress." *The Journal of Negro Education*, 63-82.

[165] Gebhardt, J. (2019). "Microaggressions in the workplace." SurveyMonkey: Curiosity at Work. https://www.surveymonkey.com/curiosity/microaggressions-research/.

On the individualism end, the basic unit of survival is the individual:

- The needs of the individual are satisfied over the group.
- Autonomy, independence, and self-reliance are valued.
- Individual effort and production are sought after and praised.
- Individuals strive to accomplish their own goals, achievements, and accolades.
- Individuals understand and communicate personal wants and needs.
- There is an emphasis on an individual's privacy.

On the collectivist side, individuals are more reliant on group identity:

- The basic survival unit is the group.
- There is value in collaboration.
- They have a strong desire to be part of a group and value interdependence over independence.
- An individual realizes that personal sacrifices must be made to ensure the success of a group.
- The basic notion is that the success of a group translates into the success of the individual. In other words, "If we win, I win."[166]

What do these two distinctions have to do with microaggressions? Well, social interactions shape an individual's sense of self. For example, we often see ourselves through someone else's eyes and assign value based on how others see us. In other words, we use those around us to obtain information about ourselves.

Zhu and colleagues did an experiment where they monitored the brain activity of individuals from collectivistic cultures and individualistic cultures using electrodes. When the researchers asked the individuals from collectivistic cultures, such as East Asia, to talk about themselves, their prefrontal cortex, where higher-order thinking and their sense of self is processed, showed a moderate response. Similar results were seen for the individualistic cultures from Western society when asked the same questions. However, when the researchers asked the collectivistic group about the people closest to them, their prefrontal cortex lit up significantly, whereas in the people from individualistic cultures, the prefrontal cortex showed little activity on the

---

[166] Hofstede, G. (2011). "Dimensionalizing cultures: The Hofstede model in context." *Online Readings in Psychology and Culture*, 2(1): 2307-0919.

same question. This research shows that for an individual that is more collectivistic, they are very sensitive to their social environments, particularly as it pertains to their self-concept.

The researchers concluded that individuals from more collectivistic cultures may be more sensitive to their surroundings and what people say and do to them. Marginalized cultures tend to have more robust group identification. Of course, this experiment does not suggest that people from individualistic cultures do not compare themselves to others or only assign value to the individual. It is all relative.[167]

There are likely individuals in your organization who tend to be more individualistic and others that lean more towards the collectivistic side. If underrepresented groups tend to be more collectivistic, they may be more sensitive to microaggressions and other behaviors and actions impacting their sense of self.[168] These damaging actions and behaviors can impact their level of engagement and productivity, affecting the bottom line of any organization.

For example, there have been numerous occasions where underrepresented colleagues were in situations where someone was highlighting jokes and social media trends that pertained to Asians, people of Islamic faith, women, Latinx, or Black people. Even though these individuals sharing the jokes were attempting to introduce humor, it negatively affected the individuals whose identities were being mocked. It negatively affected their sense of self, and they felt devalued in the organization. These kinds of slights and snubs, whether intentional or unintentional, are being paid attention to by the mind and brain, and when a negative situation is experienced, the brain can recall the situation and use the slight or nub as a potential reason someone does not feel a sense of belonging, thus damaging their focus and productivity.

Organizational Cultures and Microaggression

Funny enough, while giving a speech on microaggressions, Pascal kept using a common term, "man." It is common in his culture in South Africa,

---

[167] Zhu, Y., & Han, S. (2008). "Cultural differences in the self: from philosophy to psychology and neuroscience." *Social and Personality Psychology Compass*, 2(5), 1799-1811.

[168] McGoldrick, Monica, Joseph Giordano, and Nydia Garcia-Preto, eds. (2005). *Ethnicity and Family Therapy*. Guilford Press.

and he heard it used in the United States by many people as well. So, Pascal was pumped up during his talk and used it freely, saying things like, "Man, that was good!" or "Man, you got it!" At the end of his talk, someone slipped him a note. The note was cordial, and it simply asked him to stop using that word as some women in the room had taken offense.

Initially, Pascal's reaction was to become defensive. He initially took it as a cultural attack. Yet when he calmed down, he thought about it. Clearly, these women were impacted by that term. There had to be some history regarding the word that he was not privy to. Pascal started to see that he needed to treat these women the way they wanted to be treated. It was on him to make them feel supported in that space going forward, so he stopped using the word "man."

If the above example happened to another individual, would their reaction have been the same as Pascal's? Maybe or maybe not. Like individuals, society and organizations handle conflict in different ways. Because of such differences, social scientists have identified three different moral cultures that relate to managing conflict and grievances: cultures of honor, dignity, and victimization. So, let us look at each culture individually to try and understand varied potential reactions to committing a microaggression.

In a **culture of honor**, there is an emphasis on status, which is the relative standing of someone above another. Preserving and aggressively defending one's reputation and the social image is commonplace. Wherever there is a hierarchical power structure, such as in a corporate setting, the lower-level employees accept unequal power distribution. Therefore, there is a greater chance this organization will exhibit characteristics of a culture of honor. Individuals with power in an honor culture may respond aggressively to things perceived to undermine their authority, even if those actions are small. In a culture of honor, people tend to walk on eggshells and try not to offend those in power for fear of retaliation. So, in a culture of honor, microaggressions may not be brought up, and individuals may choose to suffer in silence.

In a **culture of dignity**, having "thick skin" is emphasized and valued. A person's inner sense of security and worth prevails regardless of what people say or think. In a culture of dignity, individuals do not display vulnerability and may tolerate microaggressions. For example, individuals who bring up injustices may be labeled quarrelsome and damaging to workplace culture. In response, people choose to be silent or use third parties to intervene, such as a manager or the human resources department.

In a **culture of victimhood**, there may be a battle for social control, which involves condemning an action and attempting to keep the behavior or activity from happening. The word "victim" implies some sort of conflict, where there are grievances or points of tension between individuals in an environment. Moreover, "social" means relationships could be the source of anxiety.

In a victimhood culture, there are often differences in the status of individuals in the social environment. Often those who do not have power or privilege in an environment use a variety of tactics, like gathering individuals who may feel oppressed to enable collective power, using social media to be heard and build consensus, or using third parties, all in an attempt to bring about some sort of systemic change.

Many organizations, especially higher education institutions, have created websites to solicit examples of microaggressions individuals have experienced in the past few years. During the period of researching and writing this book, we found these websites at Brown University, Carleton College, Columbia University, Dartmouth College, Harvard University, St. Olaf College, Swarthmore College, Willamette University, McGill University (Canada), University of Oxford (United Kingdom), and the University of Sydney (Australia).[169] Some examples of microaggressions documented on these pages are:

- A college professor in her thirties was told she looked too young to be a professor.
- A Black woman was asked whether she would allow her braid to be played with.
- Individuals referred to people with mental illness as abnormal.

These examples illustrate how collective power is utilized to amplify the voices of specific identity groups that have been victimized by microaggressions. Therefore, a characteristic of victimhood culture is recognizing, addressing, and highlighting injustices.

In victimhood culture, individuals and groups accused of being unjust (those with power and privilege) start to feel oppressed. An excellent example

---

[169] Campbell, B., & Manning, J. (2018). *The Rise of Victimhood Culture: Microaggressions, Safe Spaces, and the New Culture Wars, 1*: 265.

of this is the recent complaints and concerns many people, especially parents and guardians, are bringing forward over concepts of diversity, equity, and inclusion being taught in schools. These parents and guardians react because they believe many concepts like systemic racism and microaggressions villainize whiteness and teach their kids to be ashamed of their White identity. Many of these individuals have formed organized structures to speak out and appeal to authority figures about the indoctrination of their children, who they say are the casualties of this type of education.

The major point is, in a victimhood culture, oppression becomes the focal point, and conflict may start because different identities may claim victimhood. In a culture of victimization, being a victim becomes the "privileged" status. The only way to be heard is to be oppressed, and one's oppression starts to garner support.[170]

However, the ideal culture we must strive to create in society and within our organizations is a **culture of breakthrough**. In a culture of breakthrough:

- Individuals do not have to feel oppressed for their voices to be heard. There are structures in place to obtain feedback from employees. These can take the form of engagement surveys, exit interviews, and structured one-on-one meetings.

- There is a commitment to work together to build inclusivity and break down barriers.

- There are explicit co-created norms, values, and expectations for behavior, especially pertaining to discrimination and marginalization.

  For example, some organizations have developed civil discourse and dialogue statements and guidelines, scripted organizational pillars for respect and dignity, included intentional language in their handbooks, and included short courses for employees as part of their onboarding and continual professional development.

- Pathways are provided for anyone to tell stories, share their experiences, and work towards solutions together.

---

[170] Campbell, B., & Manning, J. (2018). *The Rise of Victimhood Culture: Microaggressions, Safe Spaces, and the New Culture Wars, 1*: 265.

For example, some organizations have contracted with trained counselors and facilitators to hear about the experiences of individuals. Others have employee resource groups that contain people of a specific background to share their perspectives and experiences.

This culture will lead to more productive, healthier, and happier individuals, benefiting society and organizations.

How to Respond as a Receiver of Microaggressions

How should one respond to microaggressions? To answer this question, let us visit a real-life scenario. Many of us are aware that hair has great significance in some cultures. Unfortunately, many corporations, schools, and athletic organizations have discriminated against others based on their hairstyles. According to the CROWN (Creating a Respectful and Open World for Natural Hair) website, "Black women are 1.5 times more likely to be sent home from the workplace because of their hair."[171]

So, let us say you are a woman who has just been sent home to change your hairstyle because your boss says, "It's just too ethnic, and it may offend some of our clients." How would those words make you feel? You would probably feel embarrassed, devalued, and like you do not fit in at your workplace, right? Exactly. So, the question is, how would you handle it effectively?

Below are the suggested steps for dealing with microaggressions:

- **Ask for clarification.** For example, use a key phrase: "I didn't get that, could you please explain that to me?"

- **Rephrase and clarify**. For example, saying "to make sure we are on the same page, you are saying…"

- **Make a value statement.** Use "I," such as: "In my experience, I've found generalizing like that can be hurtful."

---

[171] Exalt Resources: HR Solutions for Business. (2021). "The Crown Act: Why States and Municipalities Are Protecting Natural Hair." https://www.exaltresources.com/the-crown-act-why-states-and-municipalities-are-protecting-natural-hair/.

- **State your emotions.** For example, use a key phrase: "What was said made me uncomfortable because..."

- **Use humor to your advantage.** For example, use a key phrase like: "I guess you've met all the Asian people on the planet."

- **Acknowledge the intention and show its impact.** Use a key phrase like: "I recognize you were trying to be funny, but it didn't come across as intended."

- **Reframe the conversation.** For example, say something like: "Would you have said that if they were (wo)men?"[172] [173]

Preparation is the key to overcoming most things, and having key phrases, as we just listed, is a first step. However, there are other essential elements to keep in mind when neutralizing microaggressions:

- Separate the action from the individual. Address the issue by saying: "What you said made me uncomfortable. I would appreciate you avoiding jokes like that," or "What was said made me uncomfortable. I would appreciate it if jokes like that are not made in the future."

- Recognize that words such as "why" or "you" trigger defensiveness. Therefore, instead of saying, "Why would you…" say, "Help me understand…so that I…"

In the end, it is not about if you say something, but about what you say, how you say it, and what you look like when you say it. It means placing the individual you address in a less defensive and more receptive stance. Remember, the goal should be to share your feelings, open another's mind to your view, and to reduce, if not eliminate, future microaggressions.

As a follow-up, because discrimination on hair had become so prevalent, the CROWN Act was created by the Dove corporation, the CROWN Coalition, and California Senator Holly J. Mitchell. This legislation was passed unanimously to protect individuals from racial discrimination based

---

[172] Jana, T., & Mejias, A. D. (2018). *Erasing Institutional Bias: How to Create Systemic Change for Organizational Inclusion.* Berrett-Koehler Publishers, Incorporated.

[173] Goodman, D. (2011). *Promoting Diversity and Social Justice: Educating People from Privileged Groups.* New York: Routledge.

on hairstyles in California. Other states followed California's lead, but more still need to do so. Yet, it shows the importance of speaking up and sharing your experience to show others how to treat you the way you want to be treated.

How to Respond as Administrator of Microaggressions

Above we discuss how to address the situation if you are the recipient, but what if you are the person guilty of committing a microaggression, intentional or not? For example, imagine you are the boss, and you asked your employee to go home and change her hairstyle. You did not think what you were saying was offensive, but in your office, she tells you she is upset. Instead of going on the immediate defense, take a breather. Then, follow the steps below in how you respond to someone addressing this situation with you:

- **Perspective take and empathize:** Invite them to share their thoughts. Use phrases like: "Could I invite you to share your perspective?" "Could you help me understand how you saw the comment differently?" or "Could you tell me more about why this is important to you?" or "Could you give me your perspective on it?"

- **Listen:** It is essential to resist the urge to interrupt or think of a counter comment. Research shows that your brain can listen to 1.6 conversations at a time. Therefore, your brain can process what someone else is saying and a little bit of your inner voice. If your inner voice is the strongest, you cannot fully grasp what others say.[174]

- **Separate:** As they talk, try to define the issues the individual is presenting and clearly separate areas of concern that may arise as the individual speaks. Repeat these points back to them to establish understanding. Reflecting language is like holding up a

---

[174] Ted Radio Hour. 2014. *How Can We Listen Better.* NPR. August 25, 2022. https://www.npr.org/transcripts/283464243.

mirror so that the person can listen to a playback of what they are saying.

- **Reflect:** If you do not agree with everything the person says, start with the least concerning area. Be sure not to parrot what they said with universal truths: "Are you feeling unsafe? Devalued?" Instead, you might say: "I heard you express many concerns, but what I am hearing is that the comment makes you feel devalued." or "I also heard you express that you feel like leadership does not listen well to your concerns."

- **Seek common ground:** Find common ground by stating an overarching goal. For example: "I agree, communication is essential for…" or "It is difficult to feel valued when…."

- **Encourage solution-mindedness:** Perhaps ask the individual: "Can you describe a situation where you have felt communication was strong and you felt like you were heard?"

- **Agree on some action items:** Be specific about actionable steps on both sides.

- **Have a follow-up plan:** You can ask, "How can I follow up with you?"[175]

In a previous section, we talked about the importance of the culture of breakthrough where individuals feel safe expressing themselves. Again, this directly links to the brain. There is a portion of the brain called the rostral cingulate cortex. This is your own self-monitoring system. This part of the brain sends out signals if you are not, for example, following goals you have set for yourself. So, in an organization with established standards, your brain will automatically monitor your behavior against those values, and you will be less likely to commit microaggressions.[176]

---

[175] Baer, M.S. (2007). *Understanding and Dealing with Conflict in Independent Schools.*

[176] Maier, M. E., Gregorio, F. D., Muricchio, T., & Pellegrino, G. D. (2015). "Impaired rapid error monitoring but intact error signaling following rostral anterior cingulate cortex lesions in humans." *Frontiers in Human Neuroscience, 9*(339). https://doi.org/10.3389/fnhum.2015.00339.

Summary

If you are someone that has committed a microaggression, you can:

- Perspective take and empathize.
- Listen.
- Separate the different concerns.
- Reflect to understand emotions.
- Seek common ground.
- Encourage solution-mindedness.
- Agree on some action items.
- Have a follow-up plan.

If you are the recipient of a microaggression, you can:

- Ask for clarification.
- Rephrase and clarify
- Make a value statement and use "I."
- State your emotions.
- Use humor to your advantage.
- Acknowledge the intention and show its impact.
- Reframe the conversation.

Chapter 7

Fear of Stereotypes

"The problem with stereotypes is not that they are untrue, but that they are incomplete. They make one story become the only story."
—Chimamanda Ngozi Adichie

Before going into the specifics of stereotype threat, it is essential to define the term "stereotype" and understand where it comes from. Stereotypes are expectations formed about an individual or a particular group of people, whether true or not. These ideas may apply to an individual or group's physical appearance, personality, speech, dress, or cultural practices like religion and food, just to name a few. For example, you might hear others say things like, "Italians eat a lot of carbohydrates," or "Asians are good at math and science." [177]

So, where do we get these ideas or images about others?

As we have stated repeatedly in this book, individuals' perceptions come from their first-hand experiences or what others have shared through television or social media. Therefore, stereotypes are not always accurate. Moreover, they can be limiting and misleading, causing a range of emotions for the individuals or groups being stereotyped.

Have you ever been stereotyped or held a stereotypical thought about someone else? Chances are the answer is yes to both questions. Imagine yourself as the person being stereotyped in any of the following three

---

[177] Weinstein, Gerald, and Donna Mellen. (1997). "Anti-Semitism Curriculum Design." In *Teaching for Diversity and Social Justice*, ed. Maurine Adams, Lee Anne Bell, and Pat Griffin. New York: Routledge.

scenarios, and then ask yourself, "Would I need to counteract or live up to the perceived assumption of others?"

- The manager of a production company is conducting annual employee reviews. She notices that Kenny, the only African American on her team, logs in twice as many hours as his colleagues. When the manager asks Kenny why, he responds, "I want to do a good job. I want the opportunity to be promoted." Kenny's dad always told him that as a Black man, he must work harder than anyone else to get the same opportunities.

- The IT department hosts a competition at a local insurance company to challenge its employees. The goal is to create the most efficient software for their client, and the winning team will receive a bonus. When the employees choose teams, Korey, an Asian male, is highly sought after. Finally, one of the employees tells Korey, "We want you because you're Asian, so you must be smart. If you're on our team, we will win!" Korey immediately feels anxious, and he fears letting his teammates down.

- A teacher conducts a history lesson on civil rights. Then, to generate ideas for their essays, the teacher asks her students their thoughts and opinions on segregation. Tim, a Caucasian student, refuses to participate for fear of being labeled a racist.

Each of the above scenarios clearly demonstrates some form of stereotype threat. In the first example, Kenny is working harder to overcome a stereotype to be promoted. In the second example, Korey is the victim of stereotyping, and he feels pressured to perform to that belief. Finally, in the last scene, Tim is afraid of speaking up because his words might be misinterpreted, and he feels he may be more quickly labeled a racist because he is White. Think about the question asked in the first paragraph as you continue to read on and apply this information to your life.

The Brain and Stereotypes

So, what role does the brain play in stereotyping? Well, at the simplest level, it starts with neuroplasticity. As we explained earlier, neuroplasticity pertains to the brain's ability to change, adapt, reorganize, and evolve in response to the experiences we have in our lives. Neuroplasticity allows individuals to learn and remember their experiences. Neuroplasticity is very active during childhood and will enable kids to learn and remember things with greater ease.[178] Therefore, those "learned experiences" that trigger the brain's responses to information are learned early. Research has confirmed that children start to show intergroup bias from the ages of about three to five. This means they begin to evaluate one's own group (the in-group) as more favorable than the out-group.

For example, a yearlong study was conducted by Van Ausdale and Feagin where they observed three-to-five-year-old children in a daycare setting that was ethnically and racially diverse. They found that children at the daycare used racial categories to include and exclude other children from activities and negotiate power in social settings.[179] Furthermore, another study showed that children as young as five years old report being affected by stereotypes and ingroup and outgroup classifications.[180]

Around the ages of three to six, the following traits begin to develop:

- The child will form social connections with other children and form friendships.

- The child will play and explore more.

- The child will become grounded in rules for behavior and how emotions are expressed.

- The child will ask more questions.

- The child will assign a value to themselves as "good" or "bad."

- The child will establish a sense of purpose, becoming a leader or a follower.

---

[178] Rapoport, J. L., & Gogtay, N. (2008). "Brain neuroplasticity in healthy, hyperactive and psychotic children: Insights from neuroimaging." *Neuropsychopharmacology*, *33*(1): 181-197.

[179] Van Ausdale, D., & Feagin, J. R. (1996). "Using racial and ethnic concepts: The critical case of very young children." *American Sociological Review, 61*(5): 779-793. http://dx.doi.org/10.2307/2096453.

[180] Hirschfeld, L. A. (2008). "Children's developing conceptions of race." In S. M. Quintana & C. McKown (Eds.), *Handbook of Race, Racism, and the Developing Child.* 37–54. Hoboken, NJ: John Wiley & Sons.

- The child will be empathetic and trusting at this stage, typically.[181]

Also, in these years, the neuroplasticity in a child's brain impacts their self-impression and self-esteem because they start to see themselves and how they perceive the world. The primary gatekeepers of the images and information children at this stage are exposed to are media, family, and friends. This is also when the unconscious mind stores images and creates reference points for the child.

In the 1940s, African American psychologists Kenneth and Mamie Clark performed their famous "doll test." The doctors assembled Black children, ages three to seven, and used four dolls that differed only in color. Most of the Black kids preferred the lighter-skinned dolls to the darker-skinned ones. They even assigned positive attributes to the lighter-skinned dolls, classifying them as "nice" and "pretty" and referring to the Black dolls as "ugly" and "bad." The doll test was used in the *Brown v. Board of Education* case to showcase that segregation was causing psychological harm to Black children.[182] This test has since been repeated in many different cultures, and similar results have been seen.

Children are often unable to categorize things into multiple dimensions at once. When they see people in one dimension, like skin color, they tend to apply multiple dimensions onto that one characteristic, such as ability, beauty, and intelligence. Although children can attribute positive and negative attributes to different races, they do not internalize these labels for themselves. So, this shows that African American kids can still have a Eurocentric preference and a high self-concept. In addition, some kids chose the Black doll as having positive attributes, and the researchers hypothesized that it was because of their exposure to positive images of Black and African American figures.[183]

Although sensitivity to stereotypes can occur at a very young age, the encouraging news is whatever is learned can be unlearned. Neuroplasticity extends into adulthood; we may just have to work harder to counteract

---

[181] Erikson, E. (1968). *Identity: Youth and Crisis.* New York: Norton.

[182] Clark, K. B., & Clark, M. P. (1950). "Emotional factors in racial identification and preference in Negro children." *Journal of Negro Education, 19*(3): 341-350.

[183] Derman-Sparks, L., & Edwards, J. O. (2010). *Anti-Bias Education for Young Children and Ourselves* (Vol. 254). National Association for the Education of Young Children.

patterns of thinking that may have been consciously and unconsciously ingrained in us.

Stereotype Threats

When someone is in a situation where they feel like their actions will confirm a stereotype, it can impede the outcome of any activity, and this is called a stereotype threat. An individual's need to react to a stereotype threat can impact their feeling of belonging in an organization. Consequently, this can, in turn, negatively shape a person's performance, whether in school, at work, or at home.

According to Steele and Aronson, a stereotype threat is "…being at risk of confirming, as a self-characteristic, a stereotype about one's social identity."[184] What does that mean? An individual may reaffirm a stereotype applied by a co-worker through their actions, intentionally or not. For example, at the beginning of the chapter, we gave a scenario about an Asian man, Korey. His colleagues wanted him on their teams because they see Asians as intelligent, which is a stereotype. If they win because of something Korey did, then the stereotype gets confirmed.

Research has demonstrated how stereotype threat has resulted in decreased performance. For example, there have been studies around women on math tests, African Americans on standardized tests, people of low socioeconomic status, and Whites compared to Blacks and Latinx people regarding natural sports ability. The subjects were all reminded of the negative stereotypes associated with their gender, ethnicity, race, or socioeconomic status. Each study showed that the groups had impaired performances because of the negative stereotypes.[185]

Why is this? When individuals are reminded of being the target of negative stereotypes, their cognitive resources are divided, and they cannot focus as intently on the task at hand. Moreover, a sense of uncertainty is created when reflecting on the negative stereotype, and there is often a concerted effort to prove the negative stereotype wrong. However, hypervigilance can hijack

---

[184] Steele, C. M., & Aronson, J. (1995). "Stereotype threat and the intellectual test performance of African Americans." *Journal of Personality and Social Psychology*, *69*(5): 797.

[185] Steele, C. M., & Aronson, J. (1995). "Stereotype threat and the intellectual test performance of African Americans." *Journal of Personality and Social Psychology*, *69*(5): 797.

one's working memory where reason, decision-making, and behavior are regulated, and individuals become unable to focus on the task. Thus, their performance suffers.

Here are some of the ways stereotype threats can impact an individual:

- Physiological awareness: increased awareness divides cognitive resources and does not make those resources available for the task at hand

- Reduced working memory affects reason, memory retrieval, decision-making, and behavior

- Impaired self-regulation, which may cause an individual's emotions to dominate

- Decreased motivation because the individual may practice learned helplessness because they feel they will not change people's stereotyped views of who they are[186]

Reactions to stereotype threats vary. However, these are some ways individuals may react to such threats:

- **Fending off the stereotype:** The person tries to overcome the obstacles identified with a particular stereotype. For example, individuals work hard to prove that they are not like the negative stereotype, even at the expense of their psychological and mental health. Fending off the stereotype can occur in a variety of ways called covering:

  o **Advocacy-based covering:** Individuals do not speak out to support their identity group.

  o **Association-based covering:** Individuals do not feel comfortable associating with their identity groups at work.

  o **Appearance-based covering:** Individuals alter their physical appearance to fit in or belong.

---

[186] Block, C. J., Koch, S. M., Liberman, B. E., Merriweather, T. J., & Roberson, L. (2011). "Contending with stereotype threat at work: A model of long-term responses." 147. *The Counseling Psychologist, 39*(4): 570-600.

    o **Affiliation-based covering:** Individuals avoid behaviors that are connected to their identity group.[187]

- **Invigoration:** Individuals work harder and overcompensate to shed the image of a potential negative stereotype. For example, the idea that a minority must work twice as hard to earn people's respect.

- **Internal attributions:** Research finds that when individuals of racial and ethnic minority groups experience discrimination, they are likely to internalize failure, and this can result in negative wellbeing.[188]

- **Assimilation:** An individual works to display, embody, and assume characteristics and attributes of the social group that has higher status in the environment. For example, individuals may lighten their skin, straighten their hair, and speak in specific ways.

- **Disengagement:** Individuals will distance themselves from the environment and suggest that their performance in that environment does not impact their self-worth.[189] This denial may be necessary for an individual to cope with the stress and maintain a level of control.

What to Do

Everyone can be affected by stereotypes. However, for minority groups, the damage can be more impactful when they already feel disengaged or believe they do not belong. How can we all cope with stereotypes? The following is a list of ways:

- **Affirmation:** Confirm one another's positives through actions. Affirmation is about valuing a person's individuality and highlighting how they contribute favorably to a team or an organization. This

---

[187] Yoshino, K. (2007). *Covering: The Hidden Assault on Our Civil Rights.* Random House Trade Paperbacks.

[188] Williams, David R. (2018). "Stress and the mental health of populations of color: Advancing our understanding of race-related stressors." *Journal of Health and Social Behavior* 59(4): 466-485.

[189] Block, C. J., Koch, S. M., Liberman, B. E., Merriweather, T. J., & Roberson, L. (2011). "Contending with stereotype threat at work: A model of long-term responses." 1Ψ7. *The Counseling Psychologist, 39*(4), 570-600.

should not be done flippantly but strategically, especially following a notable achievement.

- **Emphasize commonalities:** It is important to highlight a person's individuality and frame praise in the context of organizational values, mission, and vision. This not only highlights what an individual brings to the table but also emphasizes common and shared values with the organization. In essence, it allows an individual to be part of something bigger than themselves, which is essential for diminishing stereotype threat.

- **Counter stereotype narratives:** Have positive depictions of individuals that are from a variety of backgrounds. It may be helpful to focus on increasing the representation of minority groups in an organization, especially in positions of authority.

- **Model a growth mindset:** An organization that values a growth mindset communicates that individuals can learn, grow, and develop in different areas, even when they make mistakes. It may be necessary for an organization to frame assessments to reinforce and normalize a growth mindset approach to evaluations.

- **Use a strengths-based approach:** Give feedback focused on growth and how individuals can utilize their strengths to grow and develop. It may be important to allow the individual receiving feedback to discuss how they can use their strengths to grow and develop. Frequent check ins, both formal and informal, may be important to help people to build the necessary skills to improve.

- **Clarify expectations:** It may be necessary for organizations to clearly articulate their criteria for assessment or evaluation. This level of transparency provides structure and allows individuals to objectively recognize the areas where they are and are not proficient. Point out resources available for individuals to grow and develop while normalizing the process of utilizing these resources.

- **Equip individuals on how to manage stress and anxiety:** There are multiple ways to handle stress, including changing the environment someone is in or equipping them with the skills and tools to cope in stressful situations. While a combination of both is needed, it may be helpful to give individuals tools to deal with their negative emotions.

- **Use narrative storytelling:** Discuss examples or personal experiences of how you or others have encountered situations and the skills and tools you used to overcome those situations. This type of narrative storytelling humanizes people and their experiences and allows individuals to recognize ways to counteract instances of stereotype threat.

As you read through the previous interventions, it may become evident they require leadership and power to cultivate an environment where people can thrive. Shared humanity demands that we acknowledge that people are not perfect and will always require some level of growth. Still, all stakeholders have a responsibility and opportunity to co-create an environment where people can improve. As stakeholders work to improve themselves and their environment, they elevate the organization's potential.

Summary

If you receive a negative stereotype about your identity group, let your actions do the talking.
- Do not internalize someone else's perception of who you are as truth.
- Do not define yourself by their opinion.

If you are a leader looking to counteract stereotype threat in your organization or on your team, remember to:
- Affirm.
- Emphasize commonalities.
- Revise any negative associations by exposing yourself and others to counter stereotype narratives.
- Model a growth mindset.
- Use a strengths-based approach.
- Clarify expectations.
- Equip individuals on how to manage stress and anxiety.
- Use narrative storytelling.

Chapter 8

Saving Face

"Once you embrace your value, talents, and strengths, it neutralizes when others think less of you."
—Rob Liano

After raising her kids, Vanessa returned to the workforce and obtained her teaching license. Regardless of how well a lesson went or the encouragement she received from her colleagues, she always felt that she was not good enough and did not have the skills to achieve at a high level. Due to her age and maybe her race or ethnicity, she felt she had not yet earned people's respect. Perhaps they thought she was there to fulfill some sort of quota. Sometimes she would go unacknowledged in the hallways, and people assumed that she did not have the skills to utilize the latest technological tools. Vanessa's suggestions and ideas would not receive the same airtime as a younger, more skilled educator in meetings, but then a similar concept to what she articulated would be proposed by another person in subsequent sessions and would gain traction and sometimes be praised. In Vanessa's heart, she knew she was competent, capable, and skilled at doing the work she was assigned, but she was riddled with doubt. As she sat in a professional development session, Vanessa learned the term "imposter syndrome" for what she was feeling.

Imposter syndrome was first described in the 1970s by psychologists Suzanne Imes and Pauline Rose Clance. They explained the syndrome as individuals expressing self-doubt about their accomplishments and abilities despite evidence to the contrary.[190] People who experience imposter syndrome may feel or believe their achievements come from luck or other external factors, such as sympathy, instead of their own skills. As a result, there is an inherent feeling of undeserving.

Many individuals with imposter syndrome fear exposure as a fraud or being seen as unworthy of their positions. Some believe they have fooled others into thinking they are competent. Feelings of inadequacy and incompetence rule this individual's thoughts. In the absence of achievement, accomplishments, or accolades, there is an overall sense among these individuals that they are less and do not belong.

As in the case of minorities, many are left to further ask the question: Do others value me because of my competence, creativity, innovative thinking, experience, and perspective, or do I just check a box?[191]

The Psychology of Imposter Syndrome

The way we experience situations is a combination of our thoughts, emotions, and the sensations we feel in our bodies. For example, when we experience imposter syndrome, our feelings are derived from core negative emotions, such as anger, disgust, fear, and sadness. Additionally, we often will feel those emotions in our physical bodies; we may experience tension, have a headache, and even feel physically weak.

Although it may not be easy to spot someone suffering from this affliction, some personality characteristics, such as perfectionists, super workers, or natural geniuses, can be attributed to those with imposter syndrome:

- **Perfectionists**
  - Perfectionists possess an internal drive to exceed expectations and avoid the worst-case scenario.

---

[190] Clance, P. R., & Imes, S. A. (1978). "The imposter phenomenon in high achieving women: Dynamics and therapeutic intervention." *Psychotherapy: Theory, Research & Practice, 15*(3): 241.

[191] Chrousos, G. P., Mentis, A. A., & Dardiotis, E. (2020). "Focusing on the Neuro-Psycho-Biological and Evolutionary Underpinnings of the Imposter Syndrome." *Frontiers in Psychology, 11*: 1553. https://doi.org/10.3389/fpsyg.2020.01553.

    o   They do not do well with uncertainty and may often engage in behaviors like micromanaging and being overly critical.[192]

- **Super Worker**
  - o   Super workers are convinced they do not measure up to their colleagues or people's perceptions. These individuals tend to work harder and push themselves to obtain feelings of competence.
  - o   They want to be seen as exceptional to conceal their feelings of mediocrity.
  - o   They find that having free time is wasteful, especially when they are not engaged with work.
  - o   Some may be in a high-level position but feel they did not truly earn it. To compensate, they often work harder and longer than others to demonstrate their value.

- **Natural genius**
  - o   Natural geniuses are individuals who were called "smart" or "geniuses" when they were young.
  - o   Their aptitude was often based on the ease at which they grasped concepts or facts.
  - o   The accolades they received were outcomes-based and not process-based.
  - o   Their worth is derived from getting things done the first time and receiving praise for the ease of solving problems.
  - o   Often, these individuals avoid challenges, especially in something they may not be good at—they have a fixed mindset and see no growth opportunity in failure.

---

[192] Boyes, A. (2020). "Don't Let Perfection Be the Enemy of Productivity." *Harvard Business Review.*

Although many of the above characteristics come across as having confidence, they are quite the opposite. These individuals suffer, often significantly, from insecurity and self-doubt.[193, 194, 195, 196]

Effects of Imposter Syndrome

Imposter syndrome is not without its emotional repercussions. Of the many that are afflicted, some people may experience and demonstrate some of the following:

- **Fear of failure:** One may experience a constant fear of losing or failing. They may be overly critical of others and themselves, make a serial comparison to colleagues, and avoid challenges, such as applying for jobs that may challenge them.

- **Short-lived success:** Successes are short-lived because of the subsequent negative emotions, like shame and guilt, stemming from potential future "outings" as a fraud.

- **Baseline of stress and anxiety:** A person may live in a constant state of stress and anxiety. As a result, they may not be entirely present in their daily lives or do not feel secure pursuing their passions.

- **Negative self-talk:** A person may undergo a self-deprecating, internal dialogue, especially in stress and anxiety. This creates a self-perpetuating cycle of low self-worth and negativity.

- **Reduced productivity and increased procrastination:** People may become less productive and procrastinate due to being overwhelmed by tasks and overworked due to their responsibilities.

---

[193] Castro, J. R., & Rice, K. G. (2003). "Perfectionism and ethnicity: Implications for depressive symptoms and self-reported academic achievement." *Cultural Diversity and Ethnic Minority Psychology, 9*(1): 64.

[194] Dudău, D. P. (2014). "The relation between perfectionism and impostor phenomenon." *Procedia-Social and Behavioral Sciences, 127*: 129-133.

[195] Young, V. (2011). *The Secret Thoughts of Successful Women: Why Capable People Suffer from the Impostor Syndrome and How to Thrive in Spite of It* (46773rd Edition ed.). Currency.

[196] Wilding, M. (2021, May 15). "5 Different Types of Imposter Syndrome (and 5 Ways to Battle Each One)." The Muse. https://www.themuse.com/advice/5-different-types-of-imposter-syndrome-and-5-ways-to-battle-each-one.

- **Difficulty with collaboration and delegation:** Collaboration and delegation may be undervalued. These individuals are driven to accomplish things on their own to demonstrate their competence or avoid failure.

- **Defensiveness and externalizing:** Individuals may feel defensive because constructive criticism can be an identity trigger. Individuals may externalize blame as a means of self-preservation.

- **Need for external validation:** Individuals may need constant external validation to increase their self-worth. However, the positive emotions stemming from the validation are often short-lived.[197]

These effects do impact the individual but also the organization. If an individual feels unworthy, no matter how it manifests or is expressed, as explained above, their performance suffers. Whenever focus is diverted from a specific task, one's ability to complete that task at the highest level is compromised. So, these effects matter and should be addressed to create a more positive and productive environment.

Amplifiers of Imposter Syndrome

Imposter syndrome is something that we have fought and continue to fight. For example, when Crystal and Pascal started Synergy Consulting Company and Pascal began speaking and consulting full-time, they often felt that they were one bad business decision, speech, or one negative consulting experience away from ruining their business and careers. Although they received numerous accolades and positive feedback, they still doubted themselves. At times, Pascal and Crystal thought it would be better to just solely focus on their careers in healthcare and education rather than exist outside of their comfort zone by working for themselves. Imposter syndrome followed them like a thief always looking to rob them of their confidence, creativity, and innovative ability. Pascal frequently compared himself to other speakers in an unhealthy way and relied on people's praise to continually fuel his self-esteem. Crystal often looked at other leaders and administrators of

---

[197] Cuncic, A. (2021, November 23). "What is imposter syndrome?" Verywell Mind. November 30, 2021. https://www.verywellmind.com/imposter-syndrome-and-social-anxiety-disorder-4156469.

small-businesses and compared her leadership and business decisions to those in similar roles. When they look back at these moments, they regret not living in the moment and enjoying it more. They were too preoccupied with a non-existent future where they struggled and failed.

The strategies that help Pascal and Crystal on their journey involve learning how to capture their thoughts and replace the negative self-talk with positivity. They continually work on being masters over their emotions rather than letting their feelings take control. Crystal and Pascal are wary of pursuing short-term gratification at the expense of long-term goals. Moreover, they are learning to delegate and say "no" to things that they take on merely to boost their self-esteem or others' perceptions. This means they have to articulate a clear purpose, vision, and decision, constructing a framework that will allow them to reject things that will not facilitate progress towards their goals. Are they perfect? *No!* However, Pascal and Crystal started to ask themselves, "What could go right if we say no to this?" rather than, "What could go wrong if we say no to this?"

As you read through the characteristics bulleted previously, you may identify with some of them or know someone who carries those traits. Yet, even though imposter syndrome crosses all racial, gender, and ethnic divides, there are groups where the imposter syndrome is amplified. Specifically, people of color experience it on a deeper level with respect to tokenism, stereotype threat, the burden of representation, familial pressures, and biculturalism.

Tokenism

Tokenism is "the practice of doing something, such as hiring a person who belongs to a minority group, only to prevent criticism and give the appearance that people are being treated fairly."[198] For example, an individual's presence can be "tokenized" when organizations place minorities in certain positions to check a box or are not as intentional about hiring people with racial or ethnically diverse backgrounds. Thinking back to Vanessa's story at the beginning of the chapter, part of her insecurity was

---

[198] Sherrer, Kara. (2018). "What Is Tokenism, and Why Does It Matter in the Workplace?" Vanderbilt University Owen Graduate School of Management. May 8, 2021 https://business.vanderbilt.edu/news/2018/02/26/tokenism-in-the-workplace/.

being sensitive to whether she was a token employee and did not have the respect of her colleagues as a result.

This may cause Vanessa to constantly fear failure, have an unhealthy level of stress and anxiety, and practice self-deprecating talk. Vanessa may have felt like she had to go above and beyond to prove she deserves the promotion or position.[199] You may be a leader reading this and asking yourself, "How do I counteract tokenism in my organization?" Here are a few things to keep in mind:

- **Having an inclusive culture**. Tokenism often highlights peoples' distinctiveness, so they may be self-conscious and hypervigilant, which can lead to isolation and feelings of loneliness. An inclusive culture can exist on different levels and can include making an attempt to connect the employee-to-employee resource groups or industry-specific networks outside the organization.

- **Having a healthy feedback culture**. A healthy feedback culture will allow you to see whether the mission and vision of your organization, especially for diversity, equity, and inclusion, aligns with the reality of what people are experiencing. Please note, it is important that you do not put the underrepresented minority on the spot and ask them to speak for their identity group because this may lead to a reinforcement of tokenism. Rather, structures that promote feedback communicate an organizational value that employees are encouraged to fully participate in.

- **Broad definition of diversity.** Having a broad definition of diversity, both visible and invisible diversity, within your organization could promote larger community involvement in inclusion efforts and enable people to recognize the benefits of an inclusive culture to the organization. Therefore, inclusion is not just for a specific identity group, but important to all stakeholders in the organization.

- **Be aware of the questions traditionally underrepresented groups are asking.** The questions many individuals from underrepresented populations will be asking when they interview or when they enter the organization are as follows (using race as an example):

---

[199] Merriam-Webster. "Token," (2021). *Merriam-Webster.com Dictionary.* May 8, 2021.
    https://www.merriam-webster.com/dictionary/tokenburde.

- ○ How many people of color are at the organization?
- ○ How many people of color are in senior or executive positions?
- ○ What is the attrition rate for people of color?
- ○ What are mentorship and sponsorship structures like?
- ○ What structures are available for professional development?
- ○ What commitments has the organization made to diversity, equity, and inclusion?[200]

These questions will highlight areas that you can focus attention and work on to improve the experiences of people of color within your organization.

Stereotype Threat

We discussed stereotype threat in the previous chapter, but this type of threat exacerbates imposter syndrome. When individuals are aware or familiar with negative stereotypes associated with their identity group, they may feel they need to work hard to avoid solidifying others' beliefs about the entire population of that group. If the individual does not have imposter syndrome initially, stereotype threat can lead to it.[201]

An example of how stereotype threat can lead to imposter syndrome can be found in academic settings. Many Latinx, African American, and American Indian faculty members explain that they fear people will assume a lack of competence in fields of academia besides multicultural studies. Therefore, many who enter other fields besides multicultural studies fear they will be labeled as unqualified, undeserving of their positions, or being subject to the assumption that they are a product of an affirmative action program. These realities lead to feelings of imposter syndrome.[202]

---

[200] Niemann, Yolanda Flores. (2003). "The psychology of tokenism: Psychosocial realities of faculty of color." *Handbook of Racial and Ethnic Minority Psychology.* 100-118.

[201] Steele, C. M., & Aronson, J. (1995). "Stereotype threat and the intellectual test performance of African Americans." *Journal of Personality and Social Psychology, 69*(5): 797.

[202] Niemann, Yolanda Flores. (2003). "The psychology of tokenism: Psychosocial realities of faculty of color." *Handbook of Racial and Ethnic Minority Psychology.* 100-118.

The Burden of Representation

The third type of amplification for minorities is called the burden of representation. This is when individuals of a marginalized group assume the responsibility to set an example for others in their group. Again, there is an inherent pressure to excel and pave the way for other individuals who share their identities, which can become extensive. Although mentoring programs and various programs for professional growth can be a positive way to help lead others to successful careers, amplification can also place immense pressure on an individual, as if all the responsibility for the success of others falls on them. This is a significant burden to bear and can cause excessive stress and anxiety.[203]

Moreover, due to the perception of being highly visible, many traditionally underrepresented individuals fear speaking about issues affecting their racial group. During social gatherings, such as after-work cocktails, athletic events, and dinners, those individuals that feel the burden of representation are often on guard and feel like they cannot be fully present. This affects their ability to form authentic relationships and connections in the workplace. There is often a conflict between their inner and professional self.[204]

Familial Pressures

A research study showed that the most popular messages parents of African Americans gave their kids, implicitly and explicitly, were messages about racial pride, self-worth, racial barriers, and egalitarianism. When discussing racial barriers, many kids of color receive the message to work twice as hard without making the same mistakes or having the same missteps

---

[203] Bias, S. (n.d.) "The Burden of Representation – Or – Jeez, you people really hate a kill-joy!" http://stacybias.net/2012/02/the-burden-of-representation-or-jeez-you-people-really-hate-a-kill-joy/.

[204] Niemann, Yolanda Flores. (2003). "The psychology of tokenism: Psychosocial realities of faculty of color." *Handbook of Racial and Ethnic Minority Psychology*. 100-118.

as people from other races. This mindset is often carried into adulthood and their professional world, leading some to have imposter syndrome.[205, 206]

Peng and Wright's longitudinal study of Asian American students revealed these individuals were mainly concerned with high expectations placed on them by their families, particularly their parents. For African Americans, perfectionism also came from familial expectations where parents outwardly expressed their desire for their kids to "do better than they did." Therefore, the education, jobs, and positions these individuals pursue are often a result of their family's desires. As a result, these individuals may desire external praise over internal satisfaction and happiness. They wish to make their families proud, and subsequently, they do not always seek out their passions.[207]

Biculturalism

Biculturalism also affects imposter syndrome and is defined as one's ability to navigate two cultural environments. This involves maintaining the values, beliefs, and ways of being of one's inherited culture while adopting or existing in the mainstream culture. On the one hand, biculturalism can refer to individuals from different countries and cultures who must navigate the two realities. An individual from two different cultures must learn to exist in both worlds. Some individuals may discover that they do not fit either culture's expectations. This will result in feeling like imposters in both environments and lead to increased stress and anxiety.

Recent evidence suggests that biculturalism is not necessarily a negative thing. Biculturalism can be advantageous. It is beneficial because of the adaptability of individuals. These individuals can present diverse ideas and concepts into their existing environment to benefit others and the organization. An analysis of 140 articles suggested that those with a bicultural identity could adapt to new environments more quickly, have increased

---

[205] Neblett, E. W., Smalls, C. P., Ford, K. R., Nguyen, H. X., & Sellers, R. M. (2009). "Racial socialization and racial identity: African American parents' messages about race as precursors to identity." *Journal of Youth and Adolescence*, *38*(2), 189-203.

[206] F. Doucet, M. Banerjee, S. Parade. (2016). "What should young Black children know about race? Parents of preschoolers, preparation for bias, and promoting egalitarianism." *Journal of Early Childhood Research*, DOI: 10.1177/1476718X16630763.

[207] Peng, S. S., & Wright, D. (1994). "Explanation of academic achievement of Asian American students." *The Journal of Educational Research*, *87*(6): 346-352.

resilience, and exhibit a broader perspective on life and circumstances. Consequently, to have individuals who can successfully navigate a bicultural reality is an asset due to the influx of creativity, perspective, and innovation they can bring to their places of work and other organizations.[208]

Imposter Syndrome and Racial Reconciliation

There is another area where imposter syndrome can become prominent, and that is when we have conversations about race. Beverly Tatum Daniel explains that many of us grow up thinking of race as a taboo topic we should not discuss, especially with people who do not share our identities, despite how close we are to them.[209] Why is this? This is a space where people feel as though perfection is required.[210] People assume they must behave a certain way and say "the right things" or risk consequences, like being characterized or labeled as a racist or losing established relationships. Consequently, many people avoid the topic altogether.

As we have learned previously, perfectionism is linked to imposter syndrome, so individuals may feel like they need to present a facade to make others feel comfortable. Moreover, many people of color feel they must diminish their experiences or perspectives. In contrast, those from the dominant culture may feel they do not have anything to contribute to the conversation or will be misunderstood or put on the defensive to ward off a negative label.

Here is the thing: the expectation of perfectionism prevents relational connection and intimacy.[211] In fact, we feel more connected when we express our imperfections. It is okay to say you do not understand or know what to say. This allows for open dialogue and meaningful conversation. We do not know everything, yet we can always learn something new and empathize with

---

[208] Chang, E. C. H., Downey, C. A., Hirsch, J. K., & Lin, N. J. (Eds.). (2016). *Positive Psychology in Racial and Ethnic Groups: Theory, Research, and Practice* (p. 339). Washington, DC: American Psychological Association.

[209] Tatum, B. D. (2017). "Why are all the Black kids sitting together in the cafeteria?: And other conversations about race." Hachette UK.

[210] Caldwell, M., & Frame, O. (2016). *Let's Get Real: Exploring Race, Class, and Gender Identities in the Classroom.* Routledge.

[211] Caldwell, M., & Frame, O. (2016). *Let's Get Real: Exploring Race, Class, and Gender Identities in the Classroom.* Routledge.

others. These opportunities will enable us to recognize our shared humanity and need for interdependence and human connection.

So, there are two possible journeys we can take together. We can take the "judgment journey" or the "breakthrough journey." If we take the judgment journey, we make no room for other voices. Our voices and perspectives drown out the voices and perspectives of people with whom we have a relationship. We lead with our biases and assumptions and blame others without reflection. Thus, we perpetuate a deficit-thinking mindset of individuals that are different from us. We live at the mercy of our emotions on the judgment journey and choose not to find common ground and shared interests. We choose to not acknowledge our shared humanity.

In contrast, on the breakthrough journey, we understand that interdependence is the key to unity. We bring our voices in the form of well-thought-out questions, and we allow the other party to have the dominant voice in the conversation. In these conversations, there should be a fundamental question each party should ask the other: "How did hearing this make you feel?" Then, dialogue with understanding the beliefs and values of the other person. Finding common ground and shared interests becomes the goal while suspending judgment.

Those on the breakthrough journey believe that the expectation of perfectionism counteracts relational connection, which negates progress. We understand that relationships are not horizontal on the breakthrough journey. We have all experienced hurts, pain, burdens, and victories that position us at different places in this story of racial reconciliation. As we work to uncover these diverse narratives and stories, we seek to understand. We believe that we can learn something, grow from discomfort, and enter a new state of being from our knowledge. In essence, we hold space for grace, understanding, compassion, and empathy. We recognize that dehumanization is at the heart of division. But, when we humanize each other, we begin to heal and elevate the possibilities.

Coping Mechanisms of Imposter Syndrome

Imposter syndrome is suffered by many regardless of race or ethnicity, but as we discussed, race can enhance the negative impact. Yet, despite having this affliction, many rise in their professions and lead productive lives.

How do they do it? There are three ways: code-switching, covering, and passing.

Code-switching is deemphasizing one's speech, appearance, expressions, and behaviors in favor of what is accepted and valued. For example, a research study showed that many African Americans code-switch to elevate their social standing, have successful interactions with law enforcement, promote their sense of professionalism, and be positioned for leadership opportunities. Even though code-switching may place an individual favorably in the eyes of the dominant group, in-group members may look at intentional changes to one's personal identity as disingenuous and may even confront the individual.[212]

Covering deals with deemphasizing a part of your identity to fit in and feel like you belong. We tend to conform to other people's expectations of us and, in so doing, cover our authentic selves—our thoughts, feelings, attitudes, ideas, and emotions. For example, when going for an interview, you choose to not list activities or groups you were part of that paint you as an activist or having a particular perspective on a social issue.[213]

Passing is where we emphasize those parts of ourselves that align well with the dominant culture or the in-group. In other words, we leverage areas of alignment with the in-group. For example, you know your colleague is a big skier, and you bring up the fact that you ski too. You attempt to align with that person over something you do that coincides with the dominant culture.[214]

However, there is a caveat to this rule. We were speaking with a friend, and she asked us whether the essence of shared humanity is finding commonalities and emphasizing similarities over differences. We explained how shared humanity underlines our basic human needs, including belonging. The essence of belonging is that you can exist in an environment where you do not have to fit in. Instead, you show up as your authentic self. An environment of belonging is one where space is made for differences.

---

[212] McCluney, C. L., Robotham, K., Lee, S., Smith, R., & Durkee, M. (2019). "The costs of code-switching." *Harvard Business Review, 15.*

[213] Yoshino, K. (2007). *Covering: The Hidden Assault on Our Civil Rights.* Random House Trade Paperbacks.

[214] Yoshino, K. (2007). *Covering: The Hidden Assault on Our Civil Rights.* Random House Trade Paperbacks.

Having the confidence and security to articulate your values and beliefs is the heart of belonging.

Therefore, in the example of the skier, you would not use the fact that you ski to overcompensate for aspects of your identity that do not belong. You would feel safe to be yourself and not fear exclusion. For example, suppose you did not know how to ski, but your colleague did and talked about it often. Within shared humanity, you would be secure in saying you did not ski and not fear their dismissal of you. This takes an environment of trust and psychological safety, discussed previously. When settings are low in trust and psychological safety, that is where many feel the need to fit in and conform.

Antifragility: Thriving Despite Imposter Syndrome

Author Nassim Taleb introduced the economic concept of antifragility. Nassim explains how when adverse circumstances happen to us, they do not necessarily linearly harm us—the accumulation or the sum of negative situations do not necessarily lead to our death or demise.

For example, if someone punches you in the arm, each punch singularly does not affect you to the point where it will lead to your death. The punches may make you weaker overall, possibly due to other situations, but you also possess the ability to recover and rebound. We are sitting here today because we have survived our failures and hardships in life. Nassim Taleb also explains three states of existence when we encounter negative situations and circumstances: fragile, robust and resilient, or antifragile.

Taleb describes "fragile" as when a person emotionally crumbles because a situation is too much for them to handle. These situations, which Taleb terms "Black Swan" events, coincide with a significant life event. They are not just an insult or a criticism. Fragility can occur when individuals do not learn from the hardships and trials in their lives. In fact, many of these individuals avoid uncertainty and unpredictability.

When people position themselves as learners, they enter a different state of being—creating emotional strength. They become resilient because they understand their potential weaknesses and vulnerabilities but acknowledge their inner ability to overcome any trials life hands them. Yet, with each passing difficulty or challenge, the learner obtains greater skills in perseverance and emotional fortitude—resiliency.

In his writings, Taleb also defines resilience as working through adverse situations and returning to a baseline or functional state. For example, individuals may experience something and keep moving forward due to their will and determination. Whatever the incident may be, it does not cripple the individual. In other words, resilience is one's ability to withstand adversity.

Taleb goes on to define antifragility. Antifragility occurs when setbacks, trials, and negative situations do not just return a person to their baseline but instead make them stronger, more resilient. For example, the human body is built to be antifragile. When one lifts fifty pounds, the muscles tear but eventually heal and become stronger. When one's power heals, it does not prepare the body to lift fifty pounds but allows it to lift fifty-five pounds—the body grows stronger because it failed. Or when a pathogen like a virus or bacteria attacks the body, the body mounts an immune response allowing it to learn the pathogen and become better able to protect itself against it the next time it enters the body.

What sets antifragility apart from resilience is that it is socially constructed. In other words, an individual cannot achieve a state of antifragility on their own. Becoming antifragile depends on a collaborative effort or interdependence. Using the same examples as in the previous paragraph, when the muscle tears after lifting weights, it becomes weak but relies on the ligaments and tendons to help stabilize it and heal. Moreover, when the body learns and remembers a pathogen, like a virus or bacteria, it depends on different cell lines to perform various functions to understand the pathogen and then attack it when it returns to reinfect the individual. The body can heal itself, but an inherent interdependence must occur—antifragility, not resilience.[215, 216]

In the story of unity, reconciliation, and shared humanity, individuals need each other; antifragility must be present. To achieve this state, Taleb says redundancy is necessary. For example, the body has two kidneys so that if one fails, the other one can assume the deficit. Also, we have two lungs, multiple brain lobes, and so on to compensate when needed. Additionally, the body has trillions of cells. The heart or pancreas, for example, are major

---

[215] Taleb, N. N. (2012). *Antifragile: Things That Gain From Disorder (Vol. 3)*. Random House.

[216] Williams, P. (2020). *Becoming Antifragile: Learning to Thrive Through Disruption, Challenge and Change*. Hambone Publishing.

organs that have thousands of similar cells. If a few cells are damaged, their failure does not prevent the pancreas from functioning because the other cells work, compensating for the loss and keeping the organ functioning.

In the context of our lives, while we may not be able to predict the circumstances that we encounter, there are certain things we take into each situation and each trial we face: our mindsets, strengths, and our relationships. If you ask someone an open-ended question, "What are your gifts or skills?" they will likely tell you some of their strengths but forget many. Those close to us can encourage us, advocate for us, and point out areas of power that we may not recognize.

Think of a job interview where you may be asked to discuss your strengths. Most go into the meeting knowing what they will say because they have had time to plan for this question. But if you placed a loved one in that interview with you, without preparation, what would happen? Most likely, your loved one could rattle off a laundry list of your strengths. Why? Because they are not burdened by your self-doubt. We need others to remind us of our positives when challenges strike to help us through difficult times.

Antifragility also covers the concept of a growth mindset versus a fixed mindset. In her book *Mindset*, Carol Dweck skillfully explains the difference between the two. A fixed mindset sees failure as something to avoid and fear. A growth mindset accepts failure as a learning opportunity and grows from the experience, inevitably becoming more vital for having failed.[217]

So, when someone criticizes you or says something against you, antifragility requires us to be BRAVE:

B: **B**ecome aware of your strengths

R: **R**elationships

A: **A**dopt a growth mindset

V: **V**alue diversity (different perspectives can allow you to grow)

E: **E**levate humanity and do not lose hope

What does this have to do with imposter syndrome? From all we have talked about, someone with imposter syndrome may feel inadequate and fail to see themselves for all their positives. They are brought down by failure and typically have a fixed mindset. Yet, humans are designed to be resilient and survive life's obstacles. Antifragility addresses the importance of resiliency and having a growth mindset where we take our blows and come

---

[217] Dweck, C. S. (2008). *Mindset: The New Psychology of Success*. Random House Digital, Inc.

back stronger. It also focuses on our need to elevate and celebrate one another while dispelling others' feelings of unworthiness. Imposter syndrome breaks an individual down, and antifragility gives individuals the power to build themselves up.

Your Positive Intelligent Brain and Imposter Syndrome

If you are an individual who relates to any of the characteristics described in the previous section, it is key to ask the question Shirzad Chamine, best-selling author of *Positive Intelligence*, asks: "To what extent are your mind and thoughts sabotaging or serving you?" There are two general voices he describes—the saboteur and the sage—and asks individuals to consider where we spend most of our time. The saboteur is connected to the "survivor" brain, and the sage is connected to the positive intelligence brain. In general, the survival brain region is linked to what is known as the "primitive brain," which includes the brain stem, the limbic system, and parts of the left brain. The survival brain is generally responsible for emotions, such as anxiety, anger, disappointment, shame, guilt, regret, and blame. The positive intelligence brain comprises the medial prefrontal cortex, ACC insula cortex, and right brain parts. It is responsible for curiosity, empathy, joy, creativity, peace, calm resolve, and gratitude. When the survival brain is activated, our focus is narrowed, and we get tunnel vision—this does not allow us to think outside the box or think through the goals we may have for a conversation. In essence, productive interactions and your presence can easily be sabotaged.[218]

Therefore, not dwelling on our negative emotions can allow us to access the positive intelligence brain and think strategically about being present.

Shirzad Chamine offers the following strategies to increase your positive intelligence and other ways to counteract imposter syndrome:

- **Weaken your saboteurs.** The most common saboteur, or the master saboteur you need to be wary of, is the judge saboteur. Many unconsciously categorize and judge people's behavior and intentions, activating the survivor's brain. This may lead to an

---

[218] Chamine, S. (2012). *Positive Intelligence: Why Only 20% of Teams and Individuals Achieve Their True Potential and How You Can Achieve Yours.* Greenleaf Book Group.

unproductive conversation. The way to weaken your saboteur is to be able to recognize it, label it, and work hard to counteract it.[219]

- **Strengthen the sage.** The way to strengthen the sage is to ask, "What could go right?" When you emerge from the survivor's brain, you can empathize, have compassion towards yourself and others, find common ground, be more solution-minded, open to a path forward, and sustain a relationship. It enables you to focus on the goals you set for yourself.[220]

- **Develop your positive intelligence brain on a neurological level.**[221] Mindfulness, prayer, and meditation allow more self-control and reduce reactivity or overthinking. You shift the balance of your neurochemicals from dopamine (looking to achieve your next goal) to serotonin (being content where you are).

Other Remedies for Imposter Syndrome

Remedies are cures for our pain. Yet, can we cure ourselves of imposter syndrome, or is it something that must be fixed at the societal level? Does it require resilience or antifragility? You, a leader, family member, or friend, can encourage change by:

- **Go beyond fairness.** A recent study indicated that individuals need fairness. If they are treated with dignity without bias, then they can discover their value or how they bring worth to the group. In other words, it is incumbent upon the group members to show an individual how and why they belong.[222]

This is done by individuals in an organization, especially supervisors, consistently tapping into individuals' expertise, knowledge, and experiences. When supervisors personally seek out employees and

---

[219] Chamine, S. (2012). *Positive Intelligence: Why Only 20% of Teams and Individuals Achieve Their True Potential and How You Can Achieve Yours*. Greenleaf Book Group.

[220] Chamine, S. (2012). *Positive Intelligence: Why Only 20% of Teams and Individuals Achieve Their True Potential and How You Can Achieve Yours*. Greenleaf Book Group.

[221] Chamine, S. (2012). *Positive Intelligence: Why Only 20% of Teams and Individuals Achieve Their True Potential and How You Can Achieve Yours*. Greenleaf Book Group.

[222] Begeny, C. T., Huo, Y. J., Smith, H. J., & Ryan, M. K. (2021). "Being treated fairly in groups is important, but not sufficient: The role of distinctive treatment in groups, and its implications for mental health." *PloS one, 16*(5): e0251871.

not just send a mass email or a survey, it demonstrates intentionality in building relationships, recognizing an individual's strengths, and recognizing or discovering what they have to offer.

- **Focus on thoughts, emotions, and presence.** When someone experiences negative emotions in social or work settings, most of the activation occurs in the head, chest, and heart regions. There is also deactivation that occurs in the upper and lower limbs. Positive scenarios lead to activation of the arms. Why is this significant? Presence! Two-thirds of communication is body language or non-verbal behavior—your presence. It is essential to not give power to your default thoughts and emotions, which may negatively affect your presence in society or the workplace.[223, 224]

- **Adopt a framework for making decisions.** A framework for making decisions can help you decide what to delegate or what informed decisions need additional time. This will allow you to provide structure in your life to support your confidence in your thought processes and build trust within yourself and your choices.[225]

- **Set smaller goals.** Pick something small to work on rather than something too big to accomplish. Large or lofty goals can produce anxiety and therefore cripple progress. Smaller goals lead to smaller victories along the journey of achieving the big dream.

- **Frame your goals in favorable terms.** Frame your goals in positive terms and in a way that states a clear purpose. When there is an articulated purpose, you can see and anticipate the desired state, creating higher motivation and a greater chance of achievement.

- **Establish goals with a purpose.** Design a future-oriented vision that involves more than you. This establishes a purpose behind what you are doing. For example, determine whose feedback is most

[223] Nummenmaa, L., Glerean, E., Hari, R., & Hietanen, J. K. (2014). "Bodily maps of emotions." *Proceedings of the National Academy of Sciences*, *111*(2): 646-651.

[224] Cuddy, A. (2015). *Presence: Bringing Your Boldest Self to Your Biggest Challenges.* Hachette UK.

[225] Auster, E. & Auster-Weiss, S. (2020). "Conquer your to-do list with this simple hack." *Harvard Business Review.* https://hbr.org/2020/08/conquer-your-to-do-list-with-this-simple-hack.

important, constructive, and valuable in helping you achieve your goals.

- **Practice a growth mindset.** It is essential to recognize that you are not a finished product—you are still growing and learning. Your failures are providing opportunities for you to grow and develop.

- **Recognize and regulate negative self-talk.** It is essential to recognize your inner critic by acknowledging the thought and speaking to it. This will quiet the negative voice and reframe or replace the narrative with positivity.

The Antifragile Organization

As we mentioned previously with regard to antifragility, organizations that have structures to learn their pain points and work to resolve them position themselves to become antifragile. An organization's fragility is often exposed when there is a major event, what Taleb called the Black Swan event. An example is the publicly brutal acts of violence against Black people in America. Those organizations that were not proactive in creating structures to foster inclusion had to pivot and adapt quickly to the low morale that resulted from the deaths of George Floyd, Breonna Taylor, Ahmaud Arbery, and many others. Organizations were resilient in that they were able to react quickly and efficiently by increasing their commitments to DEI initiatives both inside and outside their organizations. Those organizations that were antifragile were able to not react but be the model for other organizations and became more attractive to people of color due to their consistent commitments to diversity, equity, and inclusion.

As mentioned in a previous chapter, an example of an antifragile organization is PricewaterhouseCoopers (PwC). Following the deaths of Philando Castile, Alton Sterling, and five Dallas police officers, CEO Tim Ryan received a message from one of his Black employees who explained that he felt as though the company was being silent in the midst of all the racial tensions. Many Black employees explained that the lack of engagement left them feeling like they could not bring their full selves to work. Even though PwC had robust diversity and inclusion initiatives, it was evident that people throughout the organization needed to process what was going on.[226, 227] They decided to hold company-wide conversations on race called *ColorBrave: A Conversation About Race*. Tim Ryan went on to found the CEO Action for Diversity & Inclusion, a consortium of more than 2,000 CEOs and presidents who are committed to "supporting a more inclusive workplace for employees, communities, and society at large."

Following the deaths of George Floyd, Ahmaud Arbery, Breonna Taylor, and other Black individuals, PwC created an action plan that included the following:[228]

- **The creation of a staff advisory council.** This council included staff from different departments and backgrounds to analyze data and devise solutions to issues that were identified.

- **Accountability.** They laid out a strategic plan and provided an update on the goals of the plans.

- **Gave funds to social justice organizations.** These organizations included the NAACP Legal Defense and Education Fund and The Center for Policing Equity.

- **Time.** PwC committed to giving their employees one week of paid time to volunteer with nonprofits.

- **Created a fellowship program.** PwC created a two-year fellowship program where employees could engage in policy discussions on how to address racial injustice in the country. The CEO Action organizations were also given the option to participate. PwC was able to act swiftly and devise a comprehensive strategy because they had been engaged in this work and learned from their past shortcomings, a hallmark of antifragility.

[226] Jana, Tiffany, and Ashley Diaz Mejias. (2018). *Erasing Institutional Bias: How to Create Systemic Change For Organizational Inclusion.* Berrett-Koehler Publishers.

[227] McGirt, Ellen. (2017, February 1). "Tim Ryan's Awakening." Fortune, https://fortune.com/longform/pwc-diversity-tim-ryan/.

[228] Ryan, Tim. "What PwC is doing to stand up against racism." PwC. Accessed: June 13, 2022. https://www.pwc.com/us/en/about-us/newsroom/press-releases/what-pwc-is-doing-to-stand-up-against-racism.html.

Summary

For leaders to counteract imposter syndrome:

1) Go beyond fairness.
2) Recognize the amplifiers of imposter syndrome:
   - Tokenism
   - Stereotype threat
   - The burden of representation
   - Familial pressures
   - Biculturalism
3) Learn from your mistakes (growth mindset).

For individuals to counteract imposter syndrome:

1) Focus on your thoughts, emotions, and presence.
2) Adopt a framework for making decisions.
3) Set smaller goals.
4) Frame your goals in favorable terms.
5) Set goals with a purpose.
6) Practice a growth mindset.
7) Recognize and regulate negative self-talk.

Chapter 9

The Impact of Status

*"Not everything that is faced can be changed,
but nothing can be changed until it is faced."*
—James Baldwin

Being a person of color in the US is an identity that can make one vulnerable. Recently, while out to dinner with another Black colleague at a high-end steak house, Pascal and his friend were acutely aware that they were the only Black men in the room. Therefore, they were intentional with their words and actions to not draw attention to themselves. They both expressed this to one another and were surprised at feeling quite guarded and hyper-aware of their surroundings.

Even though they were hyper-aware of their surroundings, others in the room were enjoying their meals, seemingly unaware of those around them. The others were comfortable within that setting. Why did Pascal and his friend feel hyper-aware and even uncomfortable at this high-end restaurant? It was because of their identities.

There are three main parts of human identity:
- How we see ourselves
- How others see us
- How we act based on how we perceive people see us

Personal identity is what makes us unique and distinct—our likes, dislikes, passions, preferences, etc. Social identity is our sense of how people identify us based on our membership within socially constructed categories, such as race, age, sexual orientation, immigration status, gender, ability, religion, and class.[229] Society puts people into groups (social categorization) based on differences and similarities. Remember back to how our brains work. Humans like patterns and grouping similar things for ease of assessment. Patterns are the brain's tool for assessing risk and making sense of our environment. As a result of these social groupings, humans have trained their brains to make sense of the world through the lens of our social identities.[230, 231]

Yet categorization can also result in negative behaviors. How so? To illustrate this concept, imagine going into a predominantly Black or White organization where you are of the opposite race. What would your concerns be? Truth is, we may behave differently based on our past experiences or innate beliefs. We do this to limit the worst-case scenario or reduce risk. However, social categorization results in identifying "us vs. them" thought processes. Often, we make sense of our environments through the lens of our salient social identities, those social identities that we have learned to be hyper-aware of in different settings. For many individuals that enter an organization where they are underrepresented, like a Black person in a predominantly White organization, race becomes a salient social identity they are hyper aware of.

Our brains have an uncanny ability to fill in missing information. Think back to when Pascal and his friend were in the predominantly White restaurant. Since they did not know anyone who was present, they started to make sense of their environments using the social identity of race. When your brain hyper focuses on a salient identity like race, it is reminded of all the narratives surrounding race—in Pascal's case, the narratives surrounding White and Black people.

Our brains often recall narratives that occurred when emotions were heightened. We, as human beings, look at events and recall them like a highlight reel. We assess a moment at its peak (the most intense point) and remember how it ended.[232] As a result, the human brain readily stores

---

[229] YW Boston. (2020, October 26). "Your full self: Social identities and the workplace." YW Boston Blog. November, 1, 2020. https://www.ywboston.org/2020/10/your-full-self-social-identities-and-the-workplace/.

[230] McLeod, S. A. (2019, October 24). "Social identity theory." Simply Psychology, https://www.simplypsychology.org/social-identity-theory.html.

[231] Tajfel, H., Turner, J. C., Austin, W. G., & Worchel, S. (1979). "An integrative theory of intergroup conflict." *Organizational Identity: A Reader*, 56-65.

[232] Kahneman, D. (2011). *Thinking, Fast and Slow*. Macmillan.

negative information, especially when coupled with strong emotions. Research studies show increased activity in the hippocampus, amygdala, and prefrontal cortex areas when experiencing negative emotions. Epinephrine and cortisol are also found to act on the hippocampus and amygdala, which mediates the storage of adverse situations.[233] Unfortunately, when it comes to race, those situations where we experience heightened emotions are largely negative, making them easier to recall.

While in the restaurant, Pascal recalls the questions that were going through his head. They were: "I hope they don't think we are going to cause trouble," "I hope the servers don't treat us unfairly because they think we don't belong here," and "I hope the people here are not racist and won't judge us because of the color of our skin." As you can ascertain, the questions Pascal asked were based on negative situations that he or people who share his identity had likely experienced. This reiterates the point that your brain is likely to readily recall negative situations associated with race.

The previous sections in this chapter illustrate that identities create a lens through which we see the world, which can have a significant impact on the experiences of people in different environments.

Assigning Value to Social Categories

In the past, using social, legal, and political systems, humans assigned value to different social categories. Today, we have become implicitly socialized to value other social types, unintentionally reinforcing stereotypes. For example, when we go into a specific neighborhood, we may make sure our doors are locked, whereas we do not do the same in another community. When we approach a homeless individual, we may shield our kids and lock our doors. We are sending subtle messages related to those social categories.[234] When we categorize ourselves within subsets society deems important, we often self-identify with them to increase our sense of status within society. We also form generalized impressions of other particular social groups. For example, we could say that Millennials are like this, Gen-

[233] Vazdarjanova A, McGaugh JL. (1998). "Basolateral amygdala is not critical for cognitive memory of contextual fear conditioning." *Proc Natl Acad Sci USA.* 95(25):15003-15007. doi:10.1073/pnas.95.25.15003.

[234] Hardy, K. (2018). "DCF Racial Justice Summit 2018 Keynote." YouTube. https://www.youtube.com/watch?v=lRRQR6ZGCPc.

Zers are like that, and so on. A consequence of comparison is ranking and hierarchies.

Is this bad? Yes and no.

Our society is constructed to value social ranking, whether it is sports competitions, spelling bees, GPAs, or prom kings and queens.[235] As you can imagine from these examples, those who are not at the top of those hierarchies can be more sensitive to their deficiencies and inequities, attributes of a fixed mindset, whereas others, from a growth mindset, might learn from individuals who have higher GPAs and become better students.[236]

Unfortunately, sometimes hierarchies are imagined and constructed by those in power legally, socially, and politically. The attributes these individuals use to define others are often those humans have no control over—this is where "ranking" is destructive. An example is the social identity category of race. Throughout history, race has been used to assign value to humans. While we did not create this system of discrimination, we are left to deal with its consequences and repercussions.[237]

As people wrestle with the realities of race and racial injustice, they are asking themselves: how did we get here as a nation and world? "Here" means we have created a societal hierarchy based on social identities, like race. A potential answer is that people's ways of thinking were influenced and affected by people who had power, privilege, and position.

Power

Power is defined as the ability to influence people's actions or thoughts.[238] There are five primary sources of power: wealth, laws and policy, physical force, social norms, and collective power.[239] These sources enable individuals

---

[235] Koski, J. E., Xie, H., & Olson, I. R. (2015). "Understanding social hierarchies: The neural and psychological foundations of status perception." *Social Neuroscience, 10*(5): 527-550.

[236] Muscatell, K. A., Morelli, S. A., Falk, E. B., Way, B. M., Pfeifer, J. H., Galinsky, A. D., ... & Eisenberger, N. I. (2012). "Social status modulates neural activity in the mentalizing network." *Neuroimage, 60*(3): 1771-1777.

[237] Wilkerson, I. (2020). *Caste (Oprah's Book Club): The Origins of Our Discontents.* Random House.

[238] Macmillan Dictionary. "Power." Macmillan Dictionary Online. November 4, 2021. https://www.macmillandictionary.com/us/dictionary/american/power_1.

[239] Liu, E. (2014, November 4th). "How to understand power." https://www.youtube.com/watch?v=c_Eutci7ack&t=156s.

to influence other people's actions and can be seen throughout the history of our nation:

- **Wealth:** Wealth can be synonymous with control, as many wealthy Americans also influence laws and policies. Additionally, when we look at wealth in America, especially wealth passed down from generation to generation, homeownership is the most significant source.

  Toward the mid-1930s, the Federal Housing Administration (FHA) created a system for risk assessments. It did not insure mortgages in primarily African American neighborhoods. This was known as "redlining." After World War II, there was a rapid increase in suburban neighborhoods. Restrictive racial covenants were written in the deeds of many White suburban communities to restrict African Americans from living in those areas. There was a fear, amongst other things, that property values would decline and risk the integrity of the issued loans.[240] [241] In the 1950s, the realtor code of ethics prevented realtors from selling a house in White neighborhoods to a non-White family. If they did, they would be fined or even lose their license. In addition, zoning ordinances were put in place to prevent the low-income housing developments near single-family home suburbs.[242] These discriminatory actions caused many Black people to move to the inner cities. It was not until 1968 that the Fair Housing Act was passed to prohibit these actions against non-White Americans.[243]

- **Laws and Policies:** Laws and policies can influence people to act a certain way because the consequences of not adhering to those laws and policies could severely hamper their lives.

  In the 1850s, Robert Know wrote a book called *The Races of Man* that used science to indicate that people of color were intellectually

[240] Rothstein, R. (2017). *The Color of Law: A Forgotten History of How Our Government Segregated America.* Liveright Publishing.

[241] Gross, T. (2017, May 3rd). "A Forgotten History of How the US Government Segregated America." https://www.npr.org/2017/05/03/526655831/a-forgotten-history-of-how-the-u-s-government-segregated-america.

[242] Holy Post. (2020, June 14). Holy Post-Race in America. https://www.youtube.com/watch?v=AGUwcs9qJXY&t=302s.

[243] Holy Post. (2020, June 14). Holy Post-Race in America. https://www.youtube.com/watch?v=AGUwcs9qJXY&t=302s.

inferior not due to the size of their brain, but the texture of it. This work contributed to the birth of the eugenics movements, where people advocated for selective breeding. They encouraged people with "good" genes to reproduce (Whites) and discouraged the reproduction of people with "bad" genes (people of color). During the eugenics movement, there were forced sterilizations of many people of color and forced abortions. Moreover, in 28 states in 1915, marriage between "negroes," "Asians," "Indians," Latinos, and Whites was illegal. In 1967, legislation was passed that legalized interracial marriage throughout the United States.[244]

- **Physical force:** Individuals use physical power or the threat of physical force to influence the way people behave. In the past, this was done in public forms of brutality, like lynching.

Moreover, those with governmental power also controlled institutions that could exert physical force on people. For example, slave patrols were instituted during slavery, where everyday Whites were hired to surveil slaves to ensure they were not trying to escape. In some states, people who were part of the slave patrols were obligated to serve one-year terms. In addition, some states called on these slave patrols and everyday citizens to enforce slave codes that prevented slaves from leaving their plantations, gathering without White individuals being present, learning to read or write, and retaliating if approached and attacked by a White person.[245]

- **Social norms:** Social norms are considered unwritten rules, behaviors, attitudes, beliefs, and practices within a specific culture.

In order to counteract stereotype threat, many people of color have some unwritten rules that they often have to follow in order to stay safe and limit any doubt that people may have about their motives, intentions and behaviors. These norms are instituted because of the dominant narratives pertaining to people of color. Often, the identity groups that have social influence can impose what Pierre Bourdieu

---

[244] Western States Center. (2003). "Dismantling Racism: A Resource Book." Western States Center.

[245] Washington Post. "The Origins of Policing in America." September 24, 2020. Education video, https://www.youtube.com/watch?v=eBvo2OE5kqM.

calls an illusion of those who are not in power.[246] Therefore, individuals have to continually fight against those deficit narratives. Some common social norms to fight the imposed illusion placed on people of color include:

- Being mindful of your expressions and expressiveness, especially when angry, because of the threat of being labeled an "angry Black person," especially for women.

- Always keeping the receipt after a purchase in the event of being accused of stealing.

- Driving with a license and registration in an easy-to-access location in case you are pulled over. That way, you will not have to search for them and risk being considered a threat.

- Showing politeness in a predominantly white space by smiling, being relatable, and not drawing unnecessary attention to oneself so people feel safe around you.

- Making sure you have a full tank of gas when driving through rural, predominantly white areas so that you do not have to stop and risk being considered a threat.

- Ensuring that your hands are visible when walking throughout a store.

- Being mindful of having a door cracked or open during private meetings, especially if you are a male meeting with a female.

- Being weary of wearing the hood of a sweatshirt when walking.[247]

Although the list above is not exclusive, many minorities and people of color can identify with some of these social norms for behavior.

---

[246] Griffiths, Austin. (2018). "Using exploratory factor analysis and Bourdieu's concept of the illusion to examine inequality in an English school." *Power and Education* 10(1): 40-57.

[247] Sloss, Morgan. (2022). "Black People Are sharing The Rules They Follow That Most White People Don't Even Know About, And This Is So Important." *Goodful.* June 16, 2022. https://www.buzzfeed.com/morgansloss1/black-people-unwritten-rules.

- **Collective power:** When people believe in a specific cause, they can gather in large numbers to convince individuals to act differently. An example of this is the Civil Rights movement.

Over the course of history, there have been movements that have used collective power to counteract the deficit view of Blackness throughout the world. Two major movements in history were the Negritude Movement and the Harlem Renaissance. The Negritude movement occurred in the 1930s, 1940s, and 1950s and consisted of Black French-speaking and Caribbean philosophers, poets and artists who lived in Europe and were striving to affirm Blackness, shed light on the conditions Blacks found themselves in, and sought to reject the deficit view of Blackness in Francophone society. The Negritude movement provided Black people with the ability to tell their stories and shed light into their fears, hopes, and dreams. The negritude movement was inspired by the Harlem Renaissance, a literary and artistic movement that occurred in the 1920s in the United States. Black individuals expressed themselves using different artistic mediums to celebrate Blackness and be a voice of encouragement for others while being true to the struggles of the times.[248]

Privilege

Privilege refers to certain social advantages, benefits, or degrees of prestige and respect an individual receives.[249] Typically, it belongs to a specific social identity. However, we can also think of privilege as the presence of barriers protecting individuals from potential hardships, an individual's high sense of safety, or one's ability to access resources. These various forms of privilege benefited people of racial identities in the US, specifically Whites.

Whites in the US have long been able to institute laws and policies that benefited their identity group, accumulate intergenerational wealth, and significantly influence societal norms. However, it is essential to note that not all Whites benefit in every situation.

---

[248] Encyclopædia Britannica. "Negritude." Encyclopædia Britannica, inc. Accessed August 26, 2022, https://www.britannica.com/art/Negritude.

[249] García, Justin D. (2018). "Privilege (Social Inequality)." *Salem Press Encyclopedia*.

Privilege is context-dependent. There are some situations where specific identities are privileged, whereas in others, they are not. For example, even within the poorest neighborhoods, some families may be considered privileged just because they have certain advantages over others. Yet if you place those "privileged" families into another setting, they may be regarded as disadvantaged.

Privilege as a non-commodity asks a simple question: who gets to feel like they belong in society, your organization, or your institution? Think about that for a second, and then read the following statistics:

- 58% of Black professionals have encountered prejudice at work (as opposed to 15% of Whites)[250]

- 65% of Black professionals feel like they have to work harder to advance (as opposed to 16% of Whites)[251]

- 67% of Black employees say they do not have a sponsor in their organization[252]

What do the above statistics mean for our organizations? It means that, in general, White employees have an increased chance of feeling as though they belong in an organization compared to Black employees. In the studies above, belonging suggests that you are less likely to encounter prejudice and are more likely to have fewer barriers for advancement and sponsorship in an organization. In essence, a White individual may experience privilege as a non-commodity by having more opportunities to feel like they belong.

Using Wealth to Redistribute Power

It is important to recognize that those elements of power that were used to divide and marginalize can also be used to empower and unify. We can really make a difference in our organizations by wealth redistribution.

---

[250] CoQUAL Formerly, C. T. I. (2021). "Being Black in corporate America."

[251] CoQUAL Formerly, C. T. I. (2021). "Being Black in corporate America."

[252] Hancock, Bryan, Monne Williams, James Manyika, Lareina Yee, and Jackie Wong. (2021). "Race in the workplace: The Black experience in the US private sector." *New York: McKinsey.*

- **Supplier diversity:** A diverse supplier is defined as a business where 51 percent is owned and operated by traditionally marginalized and underrepresented groups.

- **Partnerships:** Leaders can form partnerships with minority councils, such as the National Minority Supplier Development Council, Women's Business Enterprise National Council, and the U.S. Hispanic Chamber of Commerce.

- **A Dedicated Diversity Officer:** Large corporations often have a position dedicated to diversity within the organization.

- **Intentional recruiting:** Organizational leaders can attend supplier diversity conferences to ensure that diverse suppliers are part of their pool when choosing vendors.[253]

What are some beneficial outcomes?

- **Economic benefits:** The National Minority Supplier Council (NMSC) indicates that Minority Business Enterprises (MBEs) produced $400 Billion in economic output and created or preserved 2.2 million jobs. MBEs also contributed $49 Billion in revenue to local, state, and federal tax authorities.[254]

- **More significant stakes:** Increased employment, more disposable income for people in historically underserved communities, and the potential for business to expand into new markets.

- **Increased competition:** Competition can improve the quality of products and reduce costs for business entities.

- **Winning the war on talent:** Individuals aware of an organization's efforts in supply chain diversity associated the brand with valuing diversity, making the brand more attractive to work for.[255]

---

[253] Bateman, A., Barrington, A., & Date, K. (2020). "Why You Need a Supplier-Diversity Program." *Harvard Business Review*. https://hbr.org/2020/08/why-you-need-a-supplier-diversity-program.

[254] National Minority Supplier Development Council. (n.d.). NMSDC Facts and Figures. https://www.nmsdc.org/Facts-and-Figures.pdf.

[255] National Minority Supplier Development Council. (n.d.). NMSDC Facts and Figures. https://www.nmsdc.org/Facts-and-Figures.pdf.

Revisiting Policies and Procedures

An organization needs to review its policies, procedures, and practices to determine whether they still serve its mission and vision. The company policies, processes, and practices must evolve with the needs of an organization, particularly when the demographics of an organization are changing. Moreover, it is essential to also make sure that policies and procedures, especially those inherited from previous administrations, comply with new laws and regulations.

There are some essential questions to ask of your policies and procedures:

- Are the intended effects of a policy and procedure being achieved?
- Are the policies and procedures still relevant?
- Are people within the organization clear on how to implement policies and procedures?
- Who is involved with reviewing and drafting various policies and practices?[256, 257, 258]

Inclusion Through Social Norms

Norms are an essential component of the culture in organizations. For example, as researchers study high-performing teams and organizations, they find a deep level of connection between people in the organization. Therefore, an essential norm in an organization is recognizing our shared humanity.

Norm #1: Shared Humanity

- Intentionally creating space to get to know people. Being intentional about understanding their experiences, beliefs, and values. In essence, understanding what they care about.

---

[256] Stryker, Nicole. (2017). "The Compliance Journey: Boosting the Value of Compliance in a Changing Regulatory Climate." KPMG. Ed. Karen Staines. KPMG LLP. Accessed July 13, 2017.

[257] Egal (n.d.). DEI) Diversity, Equity & Inclusion Checklist. https://haas.berkeley.edu/wp-content/uploads/EGAL_DEIChecklist.pdf.

[258] PowerDMS (2020, December 22) "Why it is important to review policies and procedures." https://www.powerdms.com/policy-learning-center/why-it-is-important-to-review-policies-and-procedures.

- Treating people with respect and dignity despite differences so that a future relationship can be maintained, if possible.

- Quickly listening, asking questions, and finding common interests and common ground.

Norm #2: Shared Accountability[259]

- Holding each other to a high standard of behavior by creating cultures of belonging. Therefore, if you are in a meeting or conversation and something is said or done that could be considered problematic, you act in a way that upholds the standard of respect.

- Being clear as to what standard you are upholding. Having clear examples of behaviors that will enforce the standard. Being transparent.

- Garnering feedback about how people are experiencing the culture. As the mantra goes, whatever is measured is valued. Feedback enables organizations to assess how close they are to achieving the desired outcome. If there are issues, allow the individuals to be part of the solution.

Norm #3: Articulating Goals and Action Items to Achieve the Mission and Vision

- Translating the mission and vision into actionable items that people can understand. Asking "What specific actions will relate to the mission and vision?" will make strides in connecting the goals to the mission and vision of the organization.

- Identifying barriers to achieving the mission and vision. As the undesired conditions and behaviors are communicated from past experiences and perspectives, there will be a basis for creating norms that will enable the mission and vision to be realized.

---

[259] Bregman, P. (2016). "The Right Way to Hold People Accountable." *Harvard Business Review.*

Collective Power: Employee Resource Groups (ERGs)[260, 261]

Employee Resource Groups present a great opportunity for organizations to foster an increased sense of physiological safety and promote a sense of belonging for individuals from different identity groups. As individuals' identities are affirmed, they are likely to bring more of themselves to work or be more authentic and contribute to the success of the organizations. Here is a model for forming employee resource groups:

- Issue a call for individuals who might be interested in starting an ERG. Different individuals that come forward can form a leadership team (representing different ERGs).
- Create a business and ethical case for ERGs to obtain leadership buy-in.
- Establish and articulate a clear mission and vision for ERGs and how they connect to the overall mission and vision of the organization.
- Identify priority areas, draft goals in light of those priorities, and then draft a three-month action plan (during the last month, an action plan is created for another three months):
  - Action plans should include a proposed budget.
  - A leadership team (of ERG leaders) that reports to the DEI working group, diversity coordinators, or executive sponsor.
  - Secure an executive sponsor (on the senior executive team) to whom the working group reports.
  - Draft a communication plan (emails, newsletters, blogs, videos) to let employees know about the ERG.
- Communicate with organizational employees about the ERG and allow people to sign up for groups somehow (obtain contact information).

---

[260] National Business & Disability Council and the National Employer Technical Assistance Center. (2011). "A Toolkit for Establishing and Maintaining Successful Employee Resource Groups." https://www.viscardicenter.org/wpcontent/uploads/2016/09/The-Toolkit-for-Establishing-Groups.pdf.

[261] Cordivano, S. (2019). "Starting your Employee Resource Group: A Guide for Employees." https://medium.com/sarah-cordivano/starting-your-employee-resource-group-a-guide-for-employees-63855b9e25b9.

- Host a welcome event for identity-specific ERGs.
- At subsequent meetings for specific ERGs, it will be necessary to:
  - Establish a calendar so that people can attend meetings.
    - Establish the time of day, the length of meetings, and how often the group will meet.
    - If the meetings are in-person, identify a specific location and a virtual plan for individuals unable to physically make the meeting.
  - Co-create norms and a shared value system for the group.
  - Members may have leadership roles and expectations within the resource group, like creating meeting agendas, creating minutes, creating a way for members to check in with each other (especially if they miss a meeting), and tracking attendance.
    - Establish some short-term and long-term goals for the group.
    - Determine achievable timelines for the goals.
    - Identify resources that are needed by the group.
- The ERG leadership team should set up meeting times to collaborate.
  - The ERG leadership group can be tasked with resolving or thinking through issues that may arise for the ERGs (leadership buy-in, recruitment, funding, membership).
  - The ERG leadership group can also plan workshops, invite speakers, and host networking events, community service, and awareness month celebrations.
  - Determine criteria on how success can be measured.

(Psychological) Force: Employees' Mental Health

Following a global pandemic, budget cuts, change in family circumstances, labor shortages, health challenges, increased workload, and an increase in racial tensions, it is more important than ever to support people's mental health.

Research indicates that there has been a forty-two percent decline in the mental health of the global workforce.[262] The symptoms of negative mental health include anxiety, stress, depression, burnout, trauma, and PTSD.

During the pandemic, people of color suffered mightily. Due to an increased incidence of pre-existing conditions, barriers to receiving healthcare services, over representation of people of color in (low-wage) front line jobs, inadequate educational resources, and the increase of xenophobia, particularly against Asian Americans, the mental health of people of color has seen a decline.[263, 264] Here are a few things that individuals can do in the workplace:

- During one-on-one meetings, have a "health and well-being" check-in that asks specific questions to determine how the organization can better support their mental health.

- Create spaces where people can choose to participate in community building activities or process experiences with trained professionals.

- Highlight resources, both free and paid, that employees can access to fit their needs. Organizations can expand their resources and support for mental health services.

- Invest in training or professional development that will enable employees to gain the skills and tools they need to build resilience.

- Communicate clearly about anti-discrimination and harassment policies.

- Update policies and practices to allow for more flexibility in the events that can lead to stress, like childcare (school cancellations), elderly care, and illness.

---

[262] Qualtrics. (2020). "The other COVID-19 crisis: Mental health." Retrieved from https://www.qualtrics.com/blog/confronting-mental-health/.

[263] Roberts, L. M., McCluney, C. L., Thomas, E. L., & Kim, M. (2020). "How U.S Companies Can Support Employees of Color Through the Pandemic." https://hbr.org/2020/05/how-u-s-companies-can-support-employees-of-color-through-the-pandemic.

[264] Greenwood, K., & Krol, N. (8). "Ways managers can support employees' mental health." *Harvard Business Review*. https://hbr. org/2020/08/8-ways-managers-can-support-employees-mental-health.

Summary

To be part of the solution when it comes to power and privilege, leaders can do the following:

    1) Use wealth to redistribute power.

    2) Review policies and procedures.

    3) Establish organizational norms and values.

    4) Leverage collective power through employee resource groups.

    5) Intentionally pay attention to the well-being of employees.

PART III
MEETING THE NEED
FOR BELONGING

Chapter 10

The Power of Partnership

"Together we can do great things."
—Mother Theresa

W hen we think of allyship, specific people in our lives come to mind. For instance, a White boss, who served as a mentor to Pascal, would routinely ask Pascal where he saw himself in the future. Their discussions were always open and honest, and his boss made him feel welcome. Then, when Pascal was not present and opportunities arose, the boss would often recommend Pascal or suggest how he could be involved. He championed Pascal's professional development and took an interest in his success—he amplified Pascal's voice as a faithful ally.

Additionally, we have developed relationships with others who have become confidants through the years. We have shared our issues and helped one another through difficulties by showing respect and concern while providing support. What makes our relationships so unique is that during times when something affects our racial identities, these individuals take the time to ask us:

- Are you comfortable sharing how you feel?
- What is important to you?
- How can I support you?
- What does it mean to follow up with you?

These are simple but powerful questions, and because we trust these individuals, we can be open with them and, in turn, get their support in the way we need. For us, the most significant sign they care is when they ask the questions, listen, and then respond in a way that shows they heard us. This is true allyship and critical for inclusion.

The Brain and Allyship

When individuals hear, see, or experience something, they often rely on two brain systems, system one and system two. System one consists of fast, automatic, associative, and intuitive thoughts, emotions, and behaviors. This is where instinctive reactions happen, where people "feel" specific ways and react based on what their "gut" tells them. System two consists of slow, conscious, deliberate, and analytical processes. This is higher-order thinking, where people contemplate, decipher thoughts carefully, and react consciously.

To counteract and avoid discriminatory conduct, individuals must engage system two more deliberately. This is where the capacity to exercise humility can be found. It is important to note that system one thinking may be necessary because individuals can gain important insights about how they feel about a situation and why they feel that way. It goes to understanding one's frame of reference.[265]

A study done by Richard Eibach and Joyce Ehrlinger assessed Whites and ethnic minorities (students of African, Latinx, and Asian descent) about their perception of racial progress in the United States. The researchers would analyze participants' views on whether progress is being made towards a goal using two critical points of reference: the starting point and the goal one is striving towards.[266]

For example, when we consider the journey towards unity, reconciliation, and equality, if one was basing their perspective on the past with all the inequity, divisive, and discriminatory laws and policies, an individual may believe that a lot of progress has been made. However, if we look at our current state of racial inequality compared to the ideal form of racial equality, people may determine that not enough progress is being made.

For example, one may look at the following graphs and focus on very different things:

---

[265] Kahneman, D. (2011). *Thinking, Fast and Slow.* Macmillan.

[266] Eibach, R. P., & Ehrlinger, J. (2006). "'Keep your eyes on the prize': Reference points and racial differences in assessing progress toward equality." *Personality and Social Psychology Bulletin, 32*(1): 66-77.

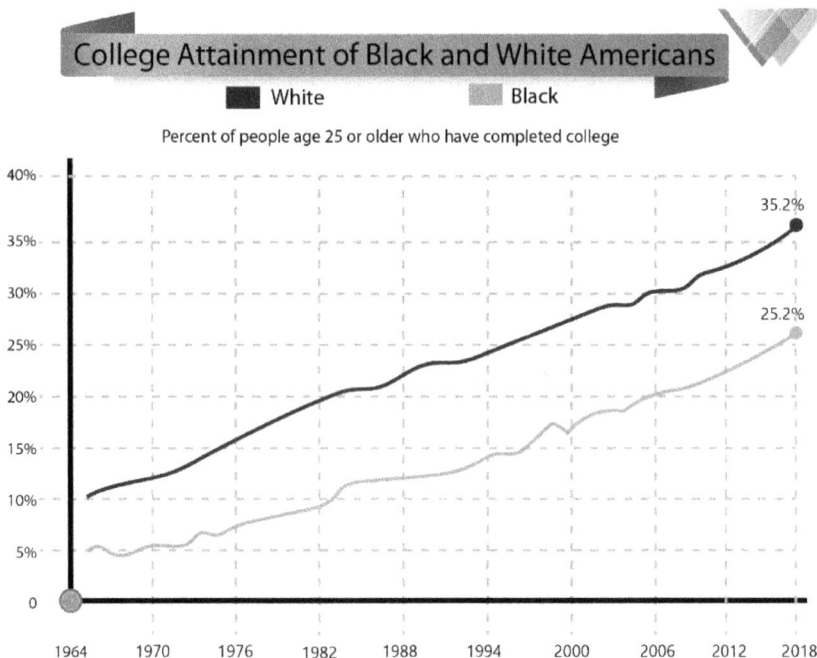

College Attainment of Black and White Americans

■ White   ■ Black

Percent of people age 25 or older who have completed college

35.2%

25.2%

Source: US Census Bureau Current Population Survey 1964 – 2018; Business Insider

Some individuals may focus on the fact that from 1964 to 2018, the college attainment of Black Americans has increased. Others may focus on the gap that still exists between Blacks and Whites. The frame of reference is different.

Let us go back to Richard Eibach and Joyce Ehrlinger's research. In that study, participants were given tasks that primed them to focus their thinking on the past racial conditions of our nation ("where we were"), the ideal state of racial equality ("where we should be prime"), or with no prime ("unframed"). The results indicated that when there was purposeful framing around the questions, both Whites and ethnic minorities tended to be on the same page.

When it comes to allyship, communication is essential. Communication means that we take time to understand each other's frames of reference in order to work together to reach a desired goal.

The Brain and Patterns

As we said previously, the brain likes ritual, predictability, and patterns. Just think of routines and patterned behavior in the morning. For example, you may wake up and immediately pour a cup of coffee. You may have a go-to mug you like using, or you may take a shower or exercise as your morning ritual. Another example is eating patterns. If you are used to eating at a particular time, you start to feel hungry when that hour hits. The predictability of routine and patterns brings comfort to you and your brain and quickly activates the brain's reward circuitry.

Why does the brain like these routines? Patterns and habits are considered safe based on past experience. Working in one's comfort zone and not having to deal with the realities and truths of racial injustice is a place of comfort for the brain. Now, imagine a change in a ritual, routine, or pattern. For example, suppose an individual becomes an ally and learns about racial injustice. The work might be energizing and invigorating at the start, but the brain longs for normalcy and predictability, which may cause the person to revert from disintegration (breaking away from one's comfort zone) to reintegration (going back into one's comfort zone). This new way of thinking and living is different and uncertain, and that is considered a threat to the brain, causing one to retreat to safety.

With a change in pattern and ritual, the brain must spend more energy (specifically the prefrontal cortex, system two) planning, analyzing, and processing the new change. Moreover, when operating in a changed and uncertain environment, individuals are more hyper-aware of their environment and continually undergo threat appraisals. For example, they may ask:

- Will I lose my circle of friends over this?
- Will the view I have of myself as a good person change?

With regular patterns broken, the brain tries to enter self-preservation mode. The brain fixates on things that can cause you harm and initiates a stress response where the brain's emotional centers begin to take over. This may cause an individual to feel overwhelmed and revert to a place of comfort.

Why is this important? If you become an ally, you need to be intentional and add structure to your days. You must develop a plan of action for your learning. You must intentionally meet new people who can mentor and encourage you when things get tough. Scheduling your days, weeks, or months and writing down your priorities will keep you motivated because your days become more predictable, and the unknowns become less of a threat. Rather than dwelling in stress and anxiety, your brain starts to focus on achieving short-term goals. If you accomplish tasks on your list, it will activate the brain's reward circuitry and cause you to feel better and think better.

So, what should you do to deal with new routines? First, make a shortlist of priorities and fight to achieve them. We will discuss goal-setting and allyship later in the chapter. But for now, know that having a structure and schedule will feed your brain's need for ritual and patterns and comfort your brain by having a level of predictability and certainty. It is all about providing "enough" comfort during the uncertainty that may accompany allyship to keep balance in your life.

Defining Allyship

An ally is an individual who advocates for unity and reconciliation alongside historically marginalized groups. In addition, allies are known for supporting empathy within their communities, institutions, and society.

An essential first step in becoming an ally is acknowledging that thoughts and behaviors (conscious and unconscious) are biased depending on one's socialization and frames of reference through which you view the world. Being an ally calls people to be committed and work consistently and intentionally. Additionally, it may provide access to spaces and places where many different backgrounds and views coexist.

Allies walk alongside historically marginalized groups and recognize they are on an educational journey, and perfectionism is not expected. This work of ally and partner is profoundly personal and emotional, but allies should expect criticism. Therefore, being an ally requires boldness, courage, and bravery, so it is critical to possess the right tools to navigate the rough waters and stay the course.

Jake Orlowitz, an activist and author of *Welcome to the Circle*, writes about allyship. Orlowitz states that if a person should make a mistake or have a misstep along the way, the best approach would be the following:

- Pause
- Listen for redirection
- Acknowledge you made a mistake
- Demonstrate humility by apologizing.

- Process and reflect on your growth
- Keep moving forward[267]

Allies are encouraged to use their voice to illuminate injustice and hold other people and themselves accountable for creating an environment of belonging. Also, they should use their voices to amplify the voices of people of color and share their stories and realities.

Yet, being an ally does not mean being invisible. An individual should never deny their personal problems or struggles. Allyship can be a two-way street, just as with our confidants. However, allyship requires an individual's problems not to take center stage at inopportune times. Allies must be mindful about drowning out the voices of those hurting and those who continue to experience feelings of unbelonging. At the right time and in the right situation, an ally's personal stories of not belonging, not feeling safe, and other experiences can allow them to connect with people who do not share their identity to foster mutual understanding and create space for interdependent learning.

Allies who are members of the majority group should enjoy the good things life offers. Shame and guilt over blessings will not help anyone. Instead, they should appreciate those gifts and use them to rejuvenate people of color. Then, they can enjoy their life and fight for equality and belonging for traditionally marginalized groups. Keeping perspective, being self-reflective, positioning as a learner, and staying committed to this noble cause is critical for success. Understand that being an ally is about creating an environment where all people feel empowered and can leave a legacy for their families.

Motivation and Allyship[268]

Allies participate in DEI work at different levels. Whether individuals do little to support DEI or are incredibly dedicated to the cause, their activity level stems from their motivation. Below are several diagrams that will explain the "zones" of motivation with allyship in DEI.

---

[267] Orlowitz, J. (2019). "17 Myths about Being a Good Ally." https://medium.com/the-j-curve/17-myths-about-being-a-good-ally-4e85413865e5.

[268] Modified from: Ryan, R. M., and Deci, E. L. (2017). *Self-Determination Theory: Basic Psychological Needs in Motivation Development and Wellness.* New York, NY: Guilford Press.

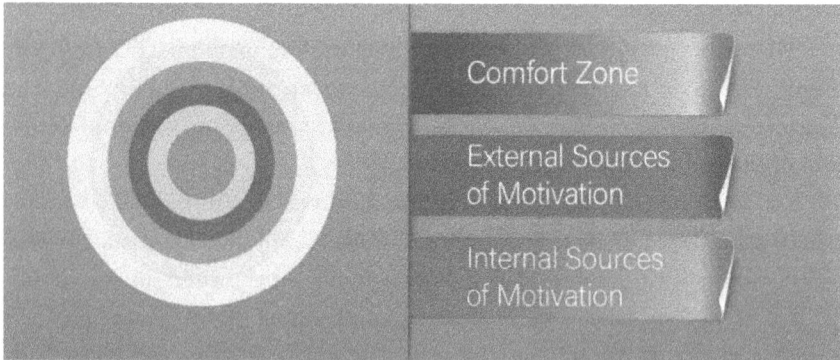

This diagram depicts the different sources of motivation that cause people to either engage in DEI or remain on the fringes. More substantial involvements are seen in those who are either external or internally motivated. Those who stay in their comfort zone are represented by the outer ring. The middle rings are the degrees of external motivation, while the inner circles (center target zone) represent internal motivation. The further one is from the target, the less motivated they may be to consistently engage in DEI work.

The above image represents the comfort zone. This may be a place where individuals passively engage with DEI work. People may have other priorities in their lives in their comfort zone and lack a sense of urgency to participate. Others may not look critically at information where their perspectives, values, beliefs, and worldviews are confirmed.

Externally motivated individuals engaged with the work of DEI, in the middle rings above, may have superficial motivations. The three reasons some are externally driven to this work are:

- Compliance
- Ego protection
- Skill enhancement

Compliance refers to going along out of fear of job loss or legal action. When individuals do DEI work for compliance or to fulfill specific requirements, the barriers to watch for are resentment, dread, apathy, cynicism, and bitterness. Remember, compliance means DEI work is being done because it is required, not necessarily because one wants to—individuals feel they are forced. These individuals are resentful because they feel that:

- Equal opportunity programs and laws give an unfair advantage to traditionally underserved groups.

- Mandatory diversity training unjustly blames Whites for the problems, issues, and barriers experienced by traditionally underserved groups.

- The organization is tokenizing specific initiatives, and their efforts will not go anywhere.

- They have been part of initiatives in the past that have not been fruitful, and they may be frustrated or unmotivated because of people's unwillingness to be more authentic in conversations relating to DEI.

Individuals who engage in this work because of compliance may not understand or recognize the benefit of this work personally or professionally. Furthermore, investing in DEI from a compliance perspective does not result in lasting change such as behavioral changes. It may, in fact, cause individuals to double down on their attitudes.[269] This is what social scientists call the backfire effect.[270]

---

[269] Kulik, C. T., Pepper, M. B., Roberson, L., & Parker, S. K. (2007). "The rich get richer: Predicting participation in voluntary diversity training." *Journal of Organizational Behavior: The International Journal of Industrial, Occupational and Organizational Psychology and Behavior, 28*(6), 753-769.

[270] Nyhan, B. (2021). "Why the backfire effect does not explain the durability of political misperceptions." *Proceedings of the National Academy of Sciences, 118*(15).

Ego protection is when externally motivated individuals get involved in DEI simply to maintain their pride and avoid criticism from others who might characterize them as insensitive, discriminatory, or prejudicial. Understand that pride is what preserves identity, social status, and others' perceptions. Ego protection can be identified by these acts:

- The individual pursues socially valued actions or tries to prevent being characterized negatively.

- The individual seeks to advertise and emphasize behaviors and actions that are socially valued.

- The individual may strive to attain praise or positive perception from individuals within or outside one's community.

For many leaders, engaging in and doing DEI work is socially valued. Those who are not engaged or adept in doing DEI work may face criticism or lose the confidence of others. Therefore, leaders may need to engage in this work to limit backlash because the cost of not doing it is too high.[271]

Skill enhancement refers to an externally motivated individual toward DEI who seeks to achieve a personal or professional goal. Typically, these individuals want to enhance their role in an organization and often compare themselves with what the market determines is essential and try to equip themselves with those skills.

Research indicates that when an individual or leader has some proficiency in concepts related to diversity and inclusion, it makes them more attractive as a candidate for jobs and more effective in the industry.[272] This may lead to cognitive dissonance. Cognitive dissonance means that an individual professing specific values and beliefs may not necessarily believe or critically think that way.[273] When an individual publicly supports a value system or convinces others they believe something, it is harder to "take it back."[274] This

---

[271] Sznycer, D., & Cohen, A. S. (2021). "How pride works." *Evolutionary Human Sciences, 3.*

[272] Bourke, J., & Espedido, A. (2019). "Why inclusive leaders are good for organizations, and how to become one." *Harvard Business Review, 3,* 2019.

[273] Cooper, J. (2019). "Cognitive dissonance: Where we've been and where we're going." *International Review of Social Psychology, 32*(1).

[274] Davis, K. E., & Jones, E. E. (1960). "Change in interpersonal perception as a means of reducing cognitive dissonance." *Journal of Abnormal and Social Psychology,* 61(3): 402–410. DOI: https://doi.org/10.1037/h0044214.

may eventually result in an individual feeling like they are not being true to themselves and sustaining negative emotions or feelings of discomfort.

Those externally motivated by skill enhancement could find themselves in an organization where the skill or expectation is not required, or an employer may not value it. As a result, the individual may drop the value or belief, defaulting to their initial state. Another danger is that their commitment to DEI work may be largely performative. Therefore, this indicates that their support for equity and inclusion will not put any meaningful action behind what is said.

It is important to note that engaging with the work from a place of compliance may cause an individual to quickly return to their comfort zone. Even so, ego protection and skill enhancement allow for greater motivation because they benefit the person. The greater the benefits of something, the more motivated one is to engage in the work. Yet, if these motivations are not solid or continuous, the individuals may abandon the work; what is not sustained retracts.

Internally motivated individuals make DEI a personal mission because they or someone they love has experienced injustice, discrimination, or marginalization. The person who is internally driven toward DEI is passionate about the work and is willing to do whatever it takes to advocate for unity and reconciliation, despite whatever opposition they face.

Internal sources of motivation have been shown to bring more lasting and long-term change compared to external motivators. Individuals internally motivated in this work have a greater understanding and appreciation and feel more responsibility. In addition, these individuals are more likely to have a strong "why." Having a strong "why" refers to having a purpose for the work. The goal then motivates allies to keep going and overcome obstacles.

Shame and Guilt

Individuals whose primary motivations are internal must guard against feelings of shame and guilt. Shame and guilt are different emotions. Shame says, "I am a bad person," and guilt says, "I did something bad." Therefore, shame speaks at the level of one's identity while guilt is more targeted towards behavior. When individuals engage with stories of struggle, hurt, and injustice, it may be easy to experience shame.[275]

Since the journey towards unity and reconciliation is personal or consistent with one's values, an individual may look at their identity, especially their race or ethnicity, and think, "I am a bad person" or "We (my identity group) are bad people." This self-condemnation or the condemnation of one's identity is often brought about by injustices being characterized as moral failures, whether the individual has directly caused the issue. Still, as a result, feelings of shame can blossom into anger and frustration with oneself and other individuals for not being committed to the work of unity, reconciliation, and social change.

An individual who persistently experiences shame may enter self-preservation mode, which may slow them down in the work leading to unity and reconciliation. Often, people who share shame avoid actions and behaviors that may exacerbate their feelings, further slowing them down. They may hear a lingering voice saying:

- You are not good enough to advocate for unity and reconciliation.
- Your people were part of the injustice.
- You are not educated enough.
- You are not worthy to say anything or even understand what people of color have gone through.

While internal motivators are not as easily corruptible, they come with emotional danger.

The vital thing to note here is that external and internal motivations enable people to be engaged. If an organization does not lead with a strong "why," people will bring their own narratives to the table, which may happen anyway. So, if organizations can ground their beliefs and culture in the "why," they will provide a purpose for their people to do the work, which will motivate them to stay engaged.

Becoming Internally Motivated

What are the critical factors for an individual to become more internally motivated and move toward that targeted center? Below are the key concepts

---

[275] Brown, B. (2012). "Listening to shame." YouTube.
https://www.youtube.com/watch?v=psN1DORYYV0&t=3s.

that an individual or organization should focus on in becoming an ally for diversity, equity, and inclusion:

- **Shared humanity:** Seek relationships and engage in conversations with individuals outside your social circles. These conversations and connections enable individuals to learn from each other and learn from their lived experiences.

- **A new understanding of history:** If someone were to go to a medical doctor or psychologist for healing, part of the protocol is to take a detailed account of the patient's history. The history provides context to help assess the root cause of the issue(s). The doctor will then treat it with medicine or behavior and lifestyle change. It is no different in allyship. Understanding the past, both the good and the bad, to learn and grow is essential to healing, unity, and reconciliation. Moving past the pain and hurt requires acknowledging what went wrong, agreements on why those wrongs occurred, forgiveness, and shared accountability.[276]

- **Humility and a growth mindset:** It is no secret that individuals may differ in their perspectives and interpretations of history. As we learned in our exploration of individual bias and the unconscious mind, our brains are constantly storing images, depictions, and characterizations of the world dependent on our families, social circles, (social) media, and the institutions we are part of. The descriptions form the frame of reference through which we view the world. As a result, two people can look at something or a piece of history and have different interpretations and perspectives. Thus, the pool of knowledge that we pull from is incomplete. To understand someone else's perspective, it is crucial to try and understand, with great humility, what the other person thinks, how they feel, and why they feel a certain way.

  Signs of humility include:
  - Being open to dialogue about ideas and alternative beliefs and values.

---

[276] Livingston, R. (2021). *The Conversation: How Seeking and Speaking the Truth about Racism Can Radically Transform Individuals and Organizations*. Currency.

    o Not taking it as a personal attack when someone contradicts a substantial value or belief.

    o The openness to bring alternative perspectives and reflect on them without immediately dismissing them.

    o Being open, where appropriate, to revise the way we think or our viewpoint in light of new information.

    o Being able to treat people with dignity, even when there is a disagreement.[277]

- **Recognizing that our anchors may be different:** The issue that many people encounter is that they have other anchors or starting points when they engage in discussions. As a result, it may be essential to be more intentional at understanding the frame of reference someone is bringing to a conversation and their personal orientation to different topics. Again, this requires us to be present in discussion and dialogue.

The "why" of allyship is to create environments with respect and dignity for one another, despite our differences. When we accomplish this seemingly Herculean task, everyone will feel as though they belong in this world, which should be enough of a motivator for anyone to become an ally in DEI.

Locus of Control

We have encountered individuals of all races and ethnicities who are skeptical of DEI work. Sometimes these individuals reject DEI because they have worked hard and achieved success in life despite many barriers. They cannot fully empathize or understand individuals seemingly dependent on others to make it in life. Often these individuals will point to examples in "rural America," particularly Whites who live below the poverty line, indicating that concepts such as privilege are incomplete or oversimplified to explain societal inequalities. These skeptics give credence to one's behavior, commitment, drive, grit, and mindset as the determining factor to their

---

[277] E.J. Krumrei-Mancuso, S.V. Rouse. (2016). "The development and validation of the comprehensive intellectual humility scale."

position in life. They believe that an individual's locus of control is their skills, abilities, capabilities, and destiny management.[278]

On the other hand, we have encountered individuals, especially from traditionally underserved and marginalized populations, who have experienced the cumulative effects of discrimination and marginalization. These people possess more of an external locus of control. They believe that social and economic constraints and barriers have had a significant impact on their destiny and future success.[279]

Therefore, those individuals with a more internal locus of control tend to blame an individual. In contrast, those with an external locus of control may be more apt to consider the environments and systemic barriers that impact one's success.

Either way, if we are preoccupied with playing the "blame game" and defending our positions on why we believe a particular perspective is proper or "fact," we lose the opportunity to understand one another and why we think the way we do. In addition, due to perceptual defense, we are more apt to not fully hear another side of an argument in favor of defending why we think a certain way and then become more sensitive to evidence that proves or justifies our point.

The conversation may be challenging when one tries to convince or move another from their perspective. However, the power to engage in a dialogue, listen well, and treat people with dignity and respect, even when we disagree, will likely invite them *not* to avoid future conversations with you. They may even be open to having a sustained relationship with you.

Understand that trust and positive emotions are entry points into compelling, complex, and courageous conversations. Shame and blame will likely limit conversations and relationships. Sharing the same beliefs and ideas is not the goal, but respectful listening and understanding should be.

---

[278] Plaut, V. C. (2010). "Diversity science: Why and how difference makes a difference." *Psychological Inquiry, 21*(2): 77-99.

[279] Shaw BA, Krause N. (2001). "Exploring race variations in aging and personal control." *The Journals of Gerontology: Series B: Psychological Sciences and Social Sciences*, 56: S119–124.

Becoming Allies for Shared Humanity

When we open the lines of communication as discussed above and honestly try to hear the other person and support one another, we can become allies. There are specific steps to being part of this shared humanity:

- **Self-knowledge:** Identify one's blind spots in this narrative of unity and reconciliation. Become self-aware of your thoughts and feelings about race and ethnicity, harmony, and reconciliation in this conversation and why you feel a certain way.

- **Self-regulation:** Control your system one thinking (emotional), the automatic responses to different worldviews and perspectives. Slow down, think about the information you are consuming, and look at it from a critical point of view—engage your higher-order thinking (system two).

- **Perspective-taking:** Consciously look at situations from the other person's perspective. That includes investing in relationships, fiction, and movie accounts of individuals' lives. Be open to understanding why individuals may feel or think a certain way. This may include understanding their beliefs and values, as well as stereotypes that characterize a particular group.

- **Skills to foster relational connection:** Work at developing humility, actively listening, conflict resolution, and collaborating skills. As with any new direction in life, planning and setting goals is the best approach.

- **Specific goals:** Ask yourself: what needs to be accomplished? Why does it need to be completed?

- **Measurable goals:** Define what constitutes success. Start by identifying and defining three milestones: a short-term win, a mid-term win, and a long-term win. How will you reward yourself for achieving your "win" at each level? Also, contemplate potential barriers you may face at each level and how you will prepare for them. Be prepared to have setbacks, address issues, and refocus on the end goal.

- **Attainable goals:** Attainability goes to creating goals within reach within a timely manner. Then break these goals down into processes that will be challenging but doable for you.

- **Relevant goals:** Being clear on the "why" of your goal is essential. Any goal must have a purpose to motivate you to achieve it. Also, remember your brain is programmed to avoid pain and find pleasure, but it will always do more to avoid pain. So, define for yourself the pain of not reaching a goal and let that drive your actions.

- **Time-bound goals:** Set timeframes within reach and be prepared to assess if you achieved the goal or not.[280, 281]

Barriers to Allyship

There are barriers to many things in life. Becoming an ally for someone or a minority group has its challenges. One must be aware of the following when moving forward toward goals:

- **Lack of awareness of the impact:** Many will fail to see the purpose of the mission.

- **The difficulty of having dialogues about race and ethnicity:** Many will shy away from conversations because they fear being judged.

- **The challenge of forming authentic relationships with individuals of different races and ethnicities:** Individuals of other races and ethnicities may be reserved or cautious about opening up to you due to their past experiences. This is where time, understanding, and persistence need to be part of your plan. It is all about building trust.

---

[280] Jagers, R. J., Rivas-Drake, D., & Borowski, T. (2018). "Equity & social and emotional learning: A cultural analysis." *CASEL Assessment Work Group Brief Series.*

[281] Robert J. Jagers, Deborah Rivas-Drake & Brittney Williams (2019). "Transformative Social and Emotional Learning (SEL): Toward SEL in Service of Educational Equity and Excellence." *Educational Psychologist*, 54(3): 162-184, DOI: 10.1080/00461520.2019.1623032.

- **The expectation of perfection:** No one expects you to say all the right things, and you will fail. Be growth-minded, accept your mistakes with grace and humility, and learn from them.

- **Lack of knowledge of how to contribute:** You will be unaware of everything that needs to be done. Go into this mission with the growth mindset to learn before you act.[282]

Understand that being an ally for another or group of individuals is not an easy road. Allies will stumble, but the goal is worth the journey for everyone to live in a society where all feel as though they belong.

Stages of Allyship[283, 284]

This section addresses different stages of beliefs encountered on an allyship journey:

- **Colorblind:** People who subscribe to "colorblindness" believe everyone should be treated equally regardless of race, culture, and ethnicity. The fundamental belief here is that society affords everyone equal opportunity and the ability for everyone to reach their full potential in life. However, in a colorblind society, individuals may not be privy to the injustices, acts of discrimination, and disadvantages that different races experience daily. There is a tendency to blame an individual for their social position in these instances.

  Individuals who subscribe to colorblindness may shift their perspective when they become more aware of people's experiences with injustice or discrimination. This occurs when they are exposed to real-world examples or learn more people's perspectives and experiences. They may also be exposed to situations where they see

---

[282] Owen, K. H. (2017). "Examining racism and white allyship among counseling psychologists."

[283] Anonymous. (n.d.). "Summary of stages of racial identity development." https://www.racialequitytools.org/resourcefiles/Compilation_of_Racial_Identity_Models_7_15_11.pdf.

[284] Helms, J. E. (1990). *Black and White racial identity: Theory, Research, and Practice.* Greenwood Press.

the law being unequally applied or read historical accounts of when it occurred to different groups.

- **Awareness:** As an individual becomes more educated about their biases and blind spots, they may start to challenge previously held views on equality, color blindness, and the discrimination and marginalization of certain groups. An individual may begin to feel shame and guilt. An individual must use these emotions productively. Otherwise, negative emotions will dominate, resulting in non-action or an individual feeling paralyzed and stuck.

When negative emotions such as guilt and shame start to dominate, an individual is likely to become defensive and enter a self-protection or self-preservation mode. However, it is essential not to stay isolated from people in the awareness stage. Instead, they must find other people advocating for unity and reconciliation, have conversations with these individuals, and get mentored by them. For example, these individuals can ask people of color how they can best support them, and as they speak, listen to them.

- **Questioning:** In the questioning stage, an individual may start to be defensive and blame the victim. Individuals try to minimize the impact and significance of certain behaviors by highlighting counterexamples to a presented reality. For example, "Look at them. They seem to be very happy and thriving." Their arguments are often based on "what they see" or read in their worlds. They usually do not want to accept or acknowledge a truth that may be uncomfortable because it can cause them to feel negative emotions such as guilt, shame, and distress.

Furthermore, it may force them to reflect on their position in a situation and alter their self-perception, which can be challenging for them to process. It is often easy to blame a marginalized individual or group of people and their beliefs, values, and worldviews. For example, you may hear statements like, "If he, she or they were more…, then…" People in this stage are likely to scrutinize actions and situations more closely, leading them to find bias-confirming information.

As a result, more effort is spent on self-preservation rather than acknowledging and solving the problem. At this stage, people will find examples to justify not being associated with behaviors. For example, you may hear things like "I am really good friends with him or her, and they feel supported by me," or "I give to a certain cause." Often, conversations are deflected, and individuals tend to bring up non-related issues to justify their views.

For people in this stage, it is crucial to recognize that the more removed you are from the work of unity and reconciliation or the more inconsistent you are in engaging in this work, the more old-world views and thought patterns that have been with you for many years emerge. Do something every day to move your education and knowledge forward. Ritualize it. Set up a daily, weekly, and monthly schedule for your education. Set up monthly appointments with people of color to learn from them and their stories.

- **Leaving space for humanity:** These individuals are primed and ready for action as they realize racial injustice and bias are real forces seeking to limit people of color. They are willing to support and affirm people of color as they work towards racial justice. People begin to find ways to use their platform and influence to support the work of unity and reconciliation. As individuals commit to learning, they start having conversations with their friends and family on different topics, such as color blindness, racial injustice, inequity, unity, and reconciliation.

They are introspective and address their unconscious biases, and they begin to understand what it means to advocate for unity and reconciliation. Individuals move from thinking of individual acts of racism to understanding where there may be systems of oppression that blindly lead individuals to act in a specific way. They are committed as allies to working with and alongside people of color in an interdependent manner.

Individuals become increasingly confident in their abilities. They can organize groups, engage in meaningful dialogue, and are action-oriented. These individuals embody unity and reconciliation in their spaces of influence. Their perspectives do not shift with the news cycles; rather, they are independent and critical thinkers.

Summary

For individuals and leaders looking to improve their level of allyship, here are a few areas to consider:

1) Relationships: In what ways are you pursuing relationships with individuals that do not share your racial or ethnic identity?

2) Identifying blind spots: In what ways are you working on understanding your blind spots so you are not leading with assumptions and creating barriers to successful dialogue?

3) Peer support: Do you have connections with other individuals committed to unity and reconciliation? If you are, are you consistently dialoguing with each other as well as encouraging each other on this allyship journey?

4) Education: Are you dedicated to educating yourself on different concepts relating to diversity and inclusion? Are you re-educating yourself on the history of our country?

5) Critical consumption: Are you paying closer attention to the media sources you consume, and are you aware of confirmation bias?

Chapter 11

How to Have Inclusive, Open, and Honest Dialogue

"When we listen and celebrate what is both common and different, we become wiser, more inclusive, and better as an organization."
—Pat Wadors

When humans feel emotionally unsafe in a space, they tend to shut down, especially when emotionally charged conversations arise. We are no different in that regard. As a matter of fact, Pascal recalls an exchange where his friend had been speaking with a group about why he believed a specific ship should not be termed a "slave ship." He thought the term was a misnomer because the ship was not solely used to transport slaves, and thus he promoted the theory that slaves were mere cargo.

Feeling his emotions rise, Pascal knew that if he were to address his words at that moment, his frustration and anger would have been talking. Pascal would not have approached the conversation using his higher-order thinking or addressed his words without attacking him. That would have resulted in an unfavorable outcome, potentially harming his friendship.

Pascal did, sometime later, speak with his friend and let him know how he disagreed with his statement. He listened and understood, and they moved past the issue. However, the critical message in this story is that Pascal knew himself well enough to know that he should calm down first and not let his emotions do the talking. The results might have been far less favorable had he addressed the statement the minute it was made.

The critical points on open dialogue, some exemplified through our experiences, that we will discuss throughout this chapter are:

- Know yourself; know your boundaries.
- Approach all conversations from your higher-order thinking, not your emotional brain.
- Find that safe space where you can speak freely.
- Have the conversation when the best outcome is likely.

As you read through this chapter, you will see that having a conversation to express yourself on racial issues is essential. Yet, it is also understandable that there will be situations where you may choose not to have a conversation. Whether you table it for later or choose to not have it at all is up to you. It all comes down to what you feel is best going forward. Each issue is situational; there is no one-size-fits-all approach, as you will see. However, we hope that this chapter will give you the tools to make those individualistic decisions.

***

Throughout this book, we have discussed how the brain works and how we are biologically designed with an intrinsic need to belong and be accepted as part of the whole. Yet, social and cultural divides have separated us throughout history. Although we as humans have progressed and come a long way in closing that gap, some divisive aspects remain, and more work needs to be done. DEI work is focused on closing such divides through learning and encouraging constructive dialogue.

Most can agree that talking about our differences and listening to one another with respect is the best approach to healing those areas that divide us. Yet many still run from the hard conversations around race. Is it because of the current cancel culture, which has ignited fear among majority populations? Is it because we are programmed to avoid anything we deem uncomfortable, such as race? Or is it because the majority cultures are unaware of what minorities have experienced? No matter the reason, we cannot ignore that DEI topics often have individuals burying their heads in the sand. However, we must acknowledge that our social environment will

217

never positively progress toward equity and inclusion if we refuse to address the tough topics or continue to deny their existence.

The Human Brain

We have explained throughout this book how the brain plays a role in how we react and deal with one another. One of the key points we have made is that the human mind is built to avoid discomfort. Basically, our brain is designed to keep us safe and free from anything that may cause us pain, emotional or physical. For example, if you are hungry, you eat. If you are tired, you sleep. If you are in fear, you will run away to find safety. Our brain is our first line of defense to promote our survival, but it also hinders our progress toward being an inclusive society. It causes us to dodge, out of self-protection, crucial conversations essential for healing those topics that divide us along racial and ethnic lines.

Even though individuals have different beliefs, values, and experiences, part of our shared humanity includes the underlying mechanisms that drive human behavior, our brains. First, our brains are great at creating associations. For example, when someone says the word "dog," a dog's picture might pop into your brain—word association. Therefore, if you associate negativity with race, you will avoid having those discussions.

As we also discussed in a previous chapter, Barbara Fredrickson developed the theory of the positivity ratio. You will need more positive experiences to overcome a negative experience, if you recall.[285] Why? The brain is more adept at remembering the negative as a means of protection. In addition, the mind will always avoid the negative before it aspires to reach a positive.[286]

The fact is that motivating others with potential positive outcomes rarely works. We can tell you how wonderful it will be once you have those difficult conversations with others about race, but the pain of those discussions registers in your brain as far more significant than the reward.

When the brain receives signals or information from the environment, it goes to the thalamus, amygdala, hippocampus, and neocortex. As we explained previously, the order in which the information reaches the different

---

[285] Fredrickson, B. L. (2013). *Updated Thinking on Positivity Ratios.*

[286] Tierney, J., & Baumeister, R. F. (2021). *The Power of Bad: How the Negativity Effect Rules Us and How We Can Rule It.* Penguin.

brain parts is why we tend toward emotional reactions first. The signal distance to the emotional brain is much shorter than the higher-order thinking portion. Therefore, we react with emotion before thought.

Looking back to the story at the beginning of this chapter, Pascal's emotions escalated long before his higher-order thinking. This is precisely why we suggest that you take a step back and take a breather when you are in an emotional state. Allow your brain time to engage the higher-order thinking for a more rational and informed reaction. Why? Because the outcome of any discussion depends on the successful transfer of information. Can you project your argument as effectively in an emotional state? No, of course not. Nobody can.

For example, American social psychologist Johnathan Haidt compared the emotional brain to an elephant and the higher-order part of the brain to the elephant's rider. If the elephant and the rider disagree, who wins? Right: the elephant. Even though the elephant can detect danger and protect you, which is positive, we must see the actual value of the rider's ability to use higher-order thinking, such as employing a strategy to yield desired results. Therefore, if the elephant is in an agreeable state, the rider's higher-order thinking can control where they go.[287, 288]

To further demonstrate this point, let us suppose your colleague makes a statement like, "You got the prime assignment! I guess that means they needed a minority to show that we are being inclusive!" He leaves the conversation with a slight chuckle, trying to pass the comment off as a joke. However, you feel incensed by his words. But if you react with emotion, will any understanding occur? Will you be able to clearly explain how his comment hurt you? No. Your anger will place your colleague on the defensive, as we have described. Your words will go unheard, and a resolution or understanding will not occur. In fact, it will most likely further the divide and create a toxic work environment.

However, you will perhaps have a more productive conversation if you approach your colleague later, after your emotional brain cools down and you have had time to develop your points. For example, suppose you approach your colleague and say, "I think you thought you were kidding earlier, but what you said hurt me. I got the assignment because I worked for it, and I'm

[287] Haidt, J. (2006). *The Happiness Hypothesis: Finding Modern Truth in Ancient Wisdom.* Basic Books.

[288] Heath, C., & Heath, D. (2011). *Switch.* Vintage Español.

qualified to do it." By not directly attacking him out of anger, you can allow him to reflect on what he said.

The other critical element of the brain dynamic is its "need to forget," or make room. Vast amounts of information hit the brain every second of every day. As a result, the brain must make space for new memories. The brain sheds nuances and stores the general gist of a conversation; the idea or overall tone of the exchange is kept along with how an individual is treated. For example, if someone speaks to you without respect or dignity, the brain breaks it down and classifies it as "bad"—storing the memory as a generality.[289]

Now, suppose you are explaining your perception to a colleague, and the other person cuts you off mid-sentence and changes the subject. That act signifies that your opinion was assigned little to no value. This may leave you irritated and hurt. Your brain will then associate that conversation with the individual as "bad" and file it away as such. Therefore, you will most likely not look forward to speaking with that individual again.

To enhance our exchanges and encourage productivity, we must note that we cannot control what the other person in a conversation will say or how they will act. Yet, there are aspects of a conversation we can control:

- What is being said—staying on topic
- The way the information is delivered—tone and physical signs, such as facial expressions and posture
- The purpose and location of the conversation—the reason for the discussion and where it takes place
- Your level of humility—applying respect and dignity to the other's point of view whether you agree or not
- Active and compassionate listening—avoiding being lost in your own thoughts when the other individual is speaking

To this last bullet point, know that when someone says something during any conversation, your mind will react to the information you are receiving. You are listening intellectually, but simultaneously, your brain judges what it hears. To limit the judgment, you should:

- **Listen to understand.** Allow someone to completely finish their thought and then check for understanding. When we listen, we allow

---

[289] Gravitz, L. (2019). "The forgotten part of memory." *Nature, 571*(7766): S12-S12.

an individual to feel heard, and then we can ask them to clarify or provide some more detailed explanations of what was being said.

- **Ask questions.** Asking questions increases the likelihood of finding common ground and allowing the individual to provide context.

Essentials of a Productive Conversation

Let us revisit the example from the prior section where your co-worker remarked about you getting a coveted assignment. Once you have cooled down and had time to think, is there a way to approach the conversation to yield positive results? Yes. The essential elements to be used for any difficult conversation will be as follows:

1. Treat others with dignity.
2. Do not focus on head talk; focus on what is said out loud.
3. Dialogue to learn values and beliefs.
4. Remember that people have different perspectives.
5. Believe that intentions are invisible.
6. Know that the blame game does not work.
7. Keep in mind that emotions connect to needs.
8. Separate the person from the behavior.
9. Choose truth over peace.
10. Focus on the experience.[290]

Treat Others With Dignity

Harvard scholar Donna Hicks explains that dignity is at the heart of having successful conversations. Treating someone as though they have inherent value and worth is crucial to influencing someone in a conversation.

There are a few hallmarks of dignity:

- **Acceptance**: We must accept we do not know all there is to know about an individual and the experiences they have had in their lives. Therefore, all conversations should be approached as a learning experience for you.

---

[290] Puiman, R. (2019). *The Mindful Guide to Conflict Resolution: How to Thoughtfully Handle Difficult Situations, Conversations, and Personalities.* Adams Media.

- **Recognition**: Recognize that conversations go beyond the words expressed. For example, people often speak about their interpretations of facts, emphasizing what is important to them and speaking according to their values. You must recognize these points to induce understanding.

- **Psychological Safety:** Create an emotional space where individuals feel they will not be shamed, ridiculed, embarrassed, marginalized, and discriminated against. It is essential to set this goal ahead of time when preparing for a difficult conversation.

- **Understanding:** It is unlikely to change someone's perspective if they do not feel heard and understood. Taking the time to know where they are coming from may give them a sense of dignity and open them to understand your perspective.[291]

It is essential to remember that dehumanization lies at the heart of discrimination and marginalization. Empathy, compassion, love, authentic listening, and acceptance humanize us. Remember that the person you are engaging with is a human being. Allowing your humanity to connect with theirs may enable you to find common ground and have a productive conversation.

Head Talk vs. What Is Said Out Loud[292]

It is essential to recognize that there is a conversation in your head and a conversation going on outside of your head in any interaction with another person.

As an example, let us imagine the following scene: Two friends, Mike and Trevor, are talking about Mike's learning journey concerning race and racial injustice. This is the actual conversation:

Trevor: "I am so glad you had a positive experience, but I think race is emphasized too much, which could be why we are still divided as a country. It would be much easier if we didn't see color."

---

[291] Hicks, D. (2019). "Reflections on Love and Dignity in Resolving Conflict." *Journal of Interreligious Studies*, (27): 67-72.

[292] Stone, D., Heen, S., & Patton, B. (2010). *Difficult Conversations: How to Discuss What Matters Most.* Penguin.

Mike: "Well, you need to see color because that is very much part of the experience of people of color. I would be careful to say that without recognizing the experience of people of color."

Trevor: "That is my perspective, and we are all entitled to our perspective and opinions."

This is the head conversation:

Trevor: I meant that by not seeing color, I do not judge anyone based on the color of their skin. I follow Martin Luther King's rule and judge someone based upon their character. Obviously, I see that he is Black, and I respect his culture. Mike is being way too sensitive.

Mike: What do you mean the race is emphasized too much? Why would you say something like that without being a person of color? When you say you don't see color, you mean you don't *see me* and my experiences. This is crazy. Is this conversation even worth my time? I don't have much energy to give to the situation. He can think whatever he needs to, but he does not get it!

The point here is that a conversation is happening simultaneously in Mike's head and Trevor's head. These thoughts occurring in each of their heads have not been overtly communicated between them, but they have impacted their connection with one another. So, we cannot necessarily control what another person says, but we can control how we respond and react to it. Perhaps if Mike had calmly explained his feelings to Trevor and vice versa, they would have found some common ground and had a more productive exchange.

Dialogue to Learn Values and Beliefs

Remember that what an individual says stems from their values, beliefs, experiences, and memories. Therefore, if you want to try and get someone to understand or see your perspective, you must be willing to play the "long game." Perspectives are learned. So, you will need to dialogue with the other person to determine where their views emanate from. If you can find a path forward with things you both care about, there is an increased likelihood of a continued and productive relationship. This may or may not happen with just one conversation, hence the term "long game." Be patient and open, and keep trying to learn about one another.[293, 294]

People Have Different Perspectives

When you are in a meaningful conversation, you may have likely sat with the topic and dissected it in many ways. The other person may not have had the same opportunity, or the issue may not hold the same value for them. Therefore, all parties may see and interpret facts from different perspectives in any given verbal exchange.

Imagine this situation: two co-workers are speaking about the diversity efforts in their organization. One of the employees, Sandra, is impressed at the recent recruitment of females and racial and ethnic minorities to the company. However, Mark has experienced and heard of his close colleagues who share his identity not being involved in decision-making and not being given the opportunities to develop their skills and advance within the organization.

This situation shows that these two individuals within the same organization bring different frames of reference to the conversation. Therefore, dialoguing for understanding is critical. A calm, well-thought-out conversation could bring both perspectives to light and allow each person to understand the other.

Intentions Are Invisible

In any given exchange, especially a difficult one, the brain is quick to ask, "What are the intentions behind this interaction?" The danger of assuming intentions is that individuals react on suppositions rather than genuinely understanding why people are saying what they are saying. When we are having a difficult or uncomfortable conversation, it is easier to define someone's intention as terrible, especially if their tone or behavior dictates it.

For example, if somebody says something hurtful, the typical reaction may be anger or sadness. When that happens, the harmed individual assumes the person intended to hurt them. Another example is if someone speaks

[293] Arbinger Institute. (2016). *The Outward Mindset: Seeing Beyond Ourselves*. Berrett-Koehler Publishers.

[294] Krupp, S., & Schoemaker, P. J. (2014). *Winning the Long Game: How Strategic Leaders Shape the Future*. Hachette UK.

over another in a meeting or does not fully listen to an individual's perspective, their intention is assumed as one to discredit or invalidate the other's viewpoint. Despite these common assumptions around intent, it is essential to be aware that choices are far more complex than the binary good and evil. Therefore, we must be mindful before assigning judgment with a binary.

Look back at the previous example of someone speaking over you. You think the person is intentionally dismissing your views. Yet, the other person actually fears that you have already misunderstood them, and their anxiety is driving their need to convey something quickly so you will not see them as "bad." Perhaps they are not trying to discriminate or marginalize but are, in fact, trying to stop you from thinking that very thing.

Another example, if you tell your co-worker, "What you said earlier is stereotyping me! I may be a Latina woman, but…"

Your co-worker interrupts with a heated, "Wow, wait. You're twisting my words!"

You think that the person is gaslighting you intentionally, but they are just trying to stop you from assuming the worst of them. Is this the best way for the conversation to go? Absolutely not, but you cannot assume your co-worker intended to silence you or treat you poorly. Assumptions are dangerous reactions.

Therefore, we suggest that you mirror the behavior you wish to see in these situations. Listen to understand and ask questions to create a deeper understanding. When you assume another's intent, you could potentially harm any future connections as your brain will label the encounter as harmful, causing you to avoid further interaction.

Blame Game Does Not Work

When we are in a difficult conversation, it may be tempting to play the "blame game." Either out of self-preservation or bias, the temptation to blame the other party or find fault with the other party may exist. When we play the blame game, we can quickly enter the drama triangle in conversations:

- Victim
- Rescuer
- Persecutor

An example is people attributing something to you in the victim role or assigning something unjustified. An automatic response would be to be defensive or distance yourself from preconceived perspectives. Unfortunately, this may result in an unproductive conversation.

As the persecutor, you may be looking to blame another individual. You may be attempting to bring something to their attention to recognize their shortcomings, amend their behavior, and have them apologize and be held accountable for what was said and done. Unfortunately, when we play the persecutor, we may be likely to attribute something to their identity rather than their behavior, which could be unproductive.

As the rescuer, you act like the people pleaser who may bring something up, but if it is received poorly, you will immediately attempt to deemphasize the impact to appease the other party. The potential problem here is that underlying issues are not addressed directly and can lead to resentment or an avoidance of a relationship. In addition, if you are the rescuer and want to solve everyone's problem, you may feel, act, and think in that order versus feel, think, and act. You may have good intentions but cause a more significant issue down the line.

We sometimes default to the different roles above when we have difficult conversations. If we get into the blame game, compassion, empathy, understanding, and solution-mindedness are compromised.[295]

Emotions Connect to Needs

One of the most important questions you can ask someone in a conversation is "How are you feeling?" or "How did that make you feel?" This will require the person to share more deeply and create a more meaningful discussion.

For example, you see a news story about a Black man being discriminated against. It makes you angry, and you ask your colleague or friend who has experienced a similar injustice what they think about it. They may answer, "It's unfair! It has to stop." The person addresses the incident. If you ask them how they feel about the news story, they might answer, "It brings up a lot of anxiety in me. I can't sleep, and I'm worried for my kids every day."

---

[295] Shmelev, I. M. (2015). "Beyond the drama triangle: The overcoming self." *Psychology. Journal of Higher School of Economics, 12*(2): 133-149.

The person has now shared their perspective and connected with you on a deeper, more meaningful level.

Remember, emotions stem from the unconscious mind. That means that feelings are strongly connected to experiences you have had in the past. The feelings that arise come from our perceptions and thoughts, and once emotions are initiated, you may not be able to move on from them. Part of the journey is recognizing that emotions are connected to your fundamental human needs. Therefore, learning how to talk about your feelings enables you to express your deeper needs.

When emotions are expressed, be on guard to avoid judgments. It is essential to recognize that if you become defensive during a conversation or work to negate someone's experience, it could be because you are not paying full attention to your own emotions and are trying to resolve them by rejecting someone's experiences. Often, debating with someone shifts the conversation from the emotional dimension to the intellectual one. Allow the other person to express their feelings fully. Ask questions to deepen your understanding before you share your feelings. Yet, when you share, make it about you and not about "proving" the other person's feelings they just shared as wrong.

Separate a Person From the Behavior

We tend to assign behaviors to an individual's character and personality. When it comes to conversations around race, there are a few adjectives we need to avoid:

- selfish
- naive
- irrational
- unethical

When we assign those labels to others, we limit our ability to hear one another on a deeper level. Instead, our conversations should set the stage for connection and understanding, not shutting out others.

Choose Truth Over Peace

One of the significant factors that many should consider is that we are all impacted by race in some way or another. Progress cannot be made when we are silenced and unable to have authentic and honest dialogues. We desire to live in a post-racial society, but we cannot do that until we have open and authentic conversations. We cannot hide from the truth, because ignoring the truth will undermine peace in the long run.

When we see ourselves as individuals rather than representing a group, we are more apt to act courageously and boldly in this racial narrative. In essence, we may be better able to recognize the humanity of the person in front of us. This liberates us from feeling like we behave on behalf of our identity group.[296]

Focus On the Experience

Your brain does not remember specifics. Instead, it retains the gist of what is said. Therefore, you often remember the tone and whether the conversation was positive or negative. You remember whether someone listened well or whether they cut you off and discredited you. How do you want people to remember the conversations you have with them? Keep that thought in your head when you dialogue with others.

Ways to Approach a Difficult Conversation

The subject of race is a difficult one. Even with all the right tools, such as what we have laid out above, there are intense conversations where you may question if there is a value in pursuing them or not. Now, we said that if we continue to avoid dialogues around tough topics such as race, we will never progress and create more inclusive environments. Therefore, it should be more about *when* and *how* than *if.* There may be situations where having no conversation is best. However, these challenging discussions deserve the extra time and consideration of whether to have them and how you should approach them if you do have them.

Now, should you decide to engage in a conversation about race, here are some points to consider:

- Be intentional: set outcome goals (What outcome do I want?)
- Prepare to influence: know what you can control (What can I control?)

---

[296] Creary, S. (2019). "Creating More Inclusive Workplaces in an Era of Discord – The Power of Helping Across Differences."
https://www.youtube.com/watch?v=DNpadtcYh5I.

- Be an active listener: listen with compassion to learn (What can I learn?)
- Practice self-mastery: know your conflict type (How will I respond?)

Intentionality

Set a desired goal such as, "I would like to understand their point better and why they feel the way they do." When you acknowledge the intention, your brain starts to self-monitor to achieve the desired outcome. Some potential goals to aspire to might be:

- To understand the person's behavior.
- To maintain a positive relationship.
- To let someone know that you were affected by something they have done or said.
- It may also be helpful to reframe how you see the individual. For example, is it a partnership that will enable you to gain mutual understanding? Is it a collaboration that will allow both parties to grow?
- Ask yourself the question: if we are divided, how do we repair the space between us? Therefore, it does not become about you and me but about how we can work together to move forward.[297]

Prepare to Influence

Preparing to influence another is all about controlling the process versus the content. Of course, the content will flow from each side, but you can control the process by how and when you have a particular exchange. Create a safe space to foster an open dialogue where all parties can contribute freely and everyone is heard.[298]

---

[297] Creary, S. (2019). "Creating More Inclusive Workplaces in an Era of Discord – The Power of Helping Across Differences."
https://www.youtube.com/watch?v=DNpadtcYh5I.

[298] Sue, Derald Wing. (2016). *Race Talk and the Conspiracy of Silence: Understanding and Facilitating Difficult Dialogues on Race.* John Wiley & Sons.

Active Listening

Active listening requires you to focus on what the other person is saying, not anticipating your comeback or the point you will make. It also entails asking questions to clarify issues and demonstrate your attention. This will also help reduce defensiveness or anxiety from the other party, as they will now feel as though they are being heard. Active listening is about keeping a conversation in the other party's world and resisting the urge to bring the conversation into yours.

Self-Mastery

Understanding how you deal with conflict and how you want to deal with the conversation is the most critical element before beginning any dialogue. In addition, you must be aware of the different conflict types a person can possess. Remember, how you present your side will directly affect the response you get.

Conflict Types

According to the Thomas-Kilmann Conflict Mode Instrument, there are five main conflict types: accommodating, avoiding, collaboration, compromising, and competing. Let us look at each kind concerning the four considerations described above:

- **Accommodating**

  The accommodating conflict type occurs when you are willing to put your own priorities aside to focus on the other person's needs.

  o **Relationship**: People who use this conflict type are often interested in preserving a relationship. There may also be a power differential, and the individual with less power feels uncomfortable challenging the other individual's point of view.

  o **Issue**: The importance of the issue discussed impacts each individual differently, or perhaps you are in the wrong, and your actions or words have impacted another individual in a significant manner.

  o **Benefits vs. consequences**: The issue may be addressed quickly, but your point of view may not be heard or appreciated in this situation. This may not matter since the problem may not be as important to them as it is to you, or there is less of a desire to preserve or grow the relationship.

o **Timing**: When there is a desire to resolve a conflict quickly, an individual does not have the energy to go through a drawn-out conflict resolution process.

- **Avoiding**

  The avoidant conflict type describes an individual who refuses to address an issue or conflict or continually pushes off having a conversation.

  o **Relationship:** People who display this conflict type are typically interested in preserving the relationship. Sometimes their emotions run high, and they believe each party may need to calm down and think through their feelings and emotions before speaking. Or, the person may suspect the conversation could be met with hostility and resentment, so they avoid the conversation. Again, this is typically done to preserve a relationship.

  o **Issue:** The issue may be necessary to the parties involved, but emotions may be high, requiring time or space for clarity and emotional regulation. Conversely, you may not care about the issue and therefore believe the discussion is not worth the time to address it.

  o **Benefits vs. consequences**: One party may characterize the avoiding party as unable to handle conflict and may lose confidence in their competence, especially when handling conflict and disagreements. But the benefit of calming down and using high-order thinking may far outweigh any initial negative perception.

  o **Timing**: A conversation may not be conducted in a short time frame, or the current environment may not allow for a free or safe space for discussion. Choosing the right time, if at all, will require some reflection.

Should you encounter someone who displays avoidance, it is vital to recognize this may be a self-preservation mode, or there may be a specific reason they do not want to converse. Perhaps the conversation may not be a priority, or there could be a deeper issue.

231

As a result, creating a safe space for individuals to share their feelings and perspectives is essential. Set the tone by letting them know that expressing their feelings and views is okay and will not be met with ire or punishment.

- **Collaboration**
  A collaborative conflict management style is one where the interests are aligned, and they are willing to meet the needs of the other. An individual looks for ways to provide a win-win solution for all members involved.
    - **Relationship**: People with a collective conflict type desire to preserve relationships.
    - **Issue**: The issue may be necessary to the parties involved, and finding a solution is key to maintaining the relationship and moving forward.
    - **Benefits vs. consequences**: For this conflict type to be effective, individuals must understand the circumstances and emotions involved. It may require time and energy, but the result could provide a positive lasting impact.
    - **Timing**: It is worth making the time and energy to invest in understanding the issues at hand. Resolving the problem for an individual, team, or organization is worth creating a safe space for discussion.

Research suggests that when we take a more collaborative approach to conversations, the medial prefrontal cortex is more engaged, meaning we are likely to humanize the other individual and understand their perspective in greater depth. Conversely, suppose we are uncompromising in our approach. In that case, the lateral prefrontal cortex is activated, enabling us to dehumanize and use people as more of a means to an end. A productive dialogue with meaningful results may go unrealized.[299]

- **Compromising**
  A compromising conflict type attempts to find partial solutions to an issue. The parties involved may or may not be satisfied with what is proposed and, therefore, are just taking a more tit-for-tat approach.

---

[299] NeuroLeadership Institute. (2021). "Power & Expectations: The Neuroscience of Group Dynamics." https://www.youtube.com/watch?v=x4JczNbUtBk.

- o **Relationship**: Relationships are meaningful, but no one leaves the conflict completely satisfied with the solution. The solution itself may be temporary, but reaching a solution is more important than getting the best solution.
- o **Issue**: The issue may be necessary to both the parties involved, and reaching a good-enough solution is more important than reaching the correct answer. This typically occurs when a discussion is at a standstill and having no solution is not an option.
- o **Benefits vs. consequences**: The solution may be unbalanced. One party gives up more than the other party, leading to resentment. Conversely, people may look at it favorably because a solution is reached rather than not having an answer.
- o **Timing**: While resolving the conflict may take time, a quicker resolution is reached than a collaborative conflict style.

- **Competing**
  This occurs when solid opinions and motives are expressed. There is an apparent conflict with a goal and purpose in mind, hence, the desire to take a firm stance to achieve those goals and desires. Often, these individuals fail to understand the perspective of others and may reject them to get their way.
  - o **Relationship**: With a competing conflict style, the outcome is more important than the relationship. Losing the connection is accepted as a potential casualty.
  - o **Issue**: The competing conflict type occurs when individuals are passionate about a topic. Often, the issue pertains to one's morals, values, beliefs, and identity.
  - o **Benefits vs. consequences**: This conflict type can be beneficial when a conflict needs to end or people need to understand or get on board with an important issue. Individuals that exhibit this style of leadership are seen as authoritarian and uncompromising.
  - o **Timing**: This may occur when a longstanding or drawn-out issue needs resolution. A competing conflict type will cause

the problem to be resolved quickly and lead to increased negative emotions.[300]

Of all the conflict types, the one that most people struggle with is the competing conflict style. When it comes to discussing issues that pertain to race and ethnicity, people often fear engaging with an individual who displays and demonstrates a competing conflict type. However, conflict related to race and ethnicity is an issue that affects the core and identity of who people are, so it is not a surprise that this conflict type may be present more frequently when resolving challenges around that topic.

No matter what conflict type you are dealing with or that you possess, there are some basic ideas to improve your communication overall when talking about DEI:

- **Use contrasting statements**.
  Contrasting statements enable the individual to build trust with the other party and clarify your intentions. Ideas take the form of don't/do statements.[301]

  An example would be an individual approaching you and saying that they have felt as though they are not being given opportunities due to their race or ethnicity. A contrasting statement would be something like:

  "I don't want you to feel devalued in this process because it takes a lot of courage to bring an issue like this forward. But, I do want to take the time to understand your situation and how you have been feeling so that we can reach a workable solution."

  The "don't" portion of the conversation is crucial because it establishes intention and addresses potential fears the individual may have about coming to you.

- **Ask someone to explain their journey**.
  Often, you receive the action part of their process when an individual approaches you. For example, there may be a lot of thoughts and experiences that have led the individual to that point. Allowing the

---

[300] Thomas, K. W. (2008). "Thomas-kilmann conflict mode." *TKI Profile and Interpretive Report*, 1.

[301] Patterson, K. (2002). *Crucial Conversations: Tools for Talking When Stakes Are High*. Tata McGraw-Hill Education.

individual to speak about how they got to a certain point will let you understand them better, find common ground, and talk to or clarify certain misconceptions.

In addition, people can also look at issues from their perspective without considering yours, so you need to share your journey or the pathway to a decision. Explaining each other's journey humanizes people in a conversation rather than looking at them as a means to an end.[302]

- **Find common ground.**
  Individuals may not see beyond their emotions in a conflict. Finding common ground and areas of connection can alleviate tension and stress. For example, if you are both parents, there could be many things you can connect over being concerned about your child's safety, education, or socialization. Once you find commonality, you will be more connected and open to hearing each other's perspectives.

- **Reframe (step out and step back in).**
  Emotions are focused and narrow. Therefore, it may be helpful for individuals to step out of the situation and reframe it into a different context. When we "reframe" an issue, we place it in a new light that better exemplifies the perspective or point.

- **Ask questions.**
  Asking clarifying questions communicates to the other party that you are listening, especially when you state a point they made and then ask a clarifying question that relates to that in some way.

- **Neutralize an aggressive approach.**
  Control the process by the following steps:
  1. Identify positive intent rather than focusing on hostile intent.
  2. Talk about their behavior, and avoid saying statements like "you did this," or "you said this," where you make the issue about the person. This will cause immediate defensiveness. Instead, address the behavior. Begin statements in the

---

[302] Patterson, K. (2002). *Crucial Conversations: Tools for Talking When Stakes Are High.* Tata McGraw-Hill Education.

following manner: "Your words…" or "Your actions…." Speak about how the behavior, not the person.

3. Propose a new approach to the conversation, remind them of the goal, and provide consequences if an agreement is not achieved.

4. Listen well, reflect on language, and find common ground.

5. Agree on an action plan.

6. End the conversation positively, if possible.

The strategies above have a common thread, or what we call the "Golden Thread." This is when we take situations out of the emotional brain and allow for more integration of higher-order thinking. Once people utilize the higher-order thinking part of their brain, their focus on emotions dissipates. This can happen when we contrast, find common ground, reframe, and ask good questions.

At the end of the day, conversations may not always look clean and pretty, so you need to establish boundaries to know when to continue or pause a conversation. Remember, it is best to talk it out if both parties are willing, open, and rational. However, when emotions are ruling the exchange, it is best to step away and discuss later or even decide that no further discussion on the topic is needed.

The Cost of Silence

We realize some will still choose to be quiet and suffer. However, we must ask, "What is the cost of such silence?" Many people of color cannot afford to be silent because it is a matter of survival for them in their workplaces. Others fear speaking up will cause even more pain or even danger for them (lost jobs, lost possibilities for advancement, even physical harm by coworkers). Yet for those who are more willing and able to participate in conversations concerning race, the cost of silence is too high.[303]

When you wonder if you could ever get to the point where talking about issues concerning DEI is possible, remember how the brain works. It loves its default settings. Therefore, the more you choose to go without speaking

---

[303] Tatum, B. D. (2017). "Why Are All the Black Kids Still Sitting Together in the Cafeteria? and Other Conversations about Race in the Twenty-First Century." *Liberal Education*, *103*(3-4).

up, the more it becomes a conditioned response and harder to override. Yet, the more you practice these conversations, the more routine they become and the more willing you are to engage in them and be comfortable having them.

We also must understand that the discussions around race do not just concern minorities. Everyone belongs in this conversation. Race touches all of us, no matter our color or ethnic background. When you are silent in the face of injustice, that "human" part of you pays the price, often in shame and guilt. You may be able to move on, but it could still affect you and limit you from forming meaningful connections with people who may change your life and allow you to see a different aspect of the world.

The second question we must ask is: "Who pays the ultimate price?" Those younger than us are paying that price. Silence causes the status quo to exist and means minimal change will occur. If we only speak about race when hot button events happen, we will always be reactionary, which may lead to people being defensive and feeling guilt or shame. Consequently, people may remain disconnected.

In speaking up, many people of color will have to relive numerous traumatic and painful experiences they have endured because of race. Many will have to fight self-doubt, shame, and fear. Many Whites will have to resist feelings of discomfort and the urge to hide behind their privileged identities. Silence may become less of an option because you may notice and become more sensitive to inequities. Without the ability to process what you see with trusted confidants or people you are with, you may rely on polarizing sources, like the media.

Summary

As you, an individual or leader, strive to have productive conversations about race, it is important to remember the following:

1) Treat others with dignity.
2) Do not focus on head talk.
3) Dialogue to learn values and beliefs.
4) Remember that people have different perspectives.
5) Believe that intentions are invisible.
6) Know that the blame game does not work.
7) Keep in mind that emotions connect to needs.

8) Separate the person from the behavior.
9) Choose truth over peace.
10) Focus on the experience.

Chapter 12

Conclusion

"In order to be great, you just have to care. You have to care about your world, community, and equality."
—Katori Hall

Our Goal

As children, sitting in our biology classrooms, we studied the concept of heliotropism. The heliotropic effect exists when a plant tracks and follows the sun. It does this because the sun is its source of energy. In essence, the sun gives the plant life to grow, develop, and reproduce itself. Similarly, in the social sciences, the heliotropic effect asserts that individuals and organizations must gravitate towards what gives them life. Diversity gives an organization life. There are numerous studies that illustrate the impact diversity can have in an organization. Therefore, diversity is an asset to your relationships and organization. For organizations, especially, a culture of inclusion and belonging needs to be the foundation on which your organization is built so that you can reap the harvest from the diverse identities within your organizations.

The saying is true: we cannot give what we do not have. Therefore, another hallmark of life is reproduction—to produce more of something. If we do not invest in turning our knowledge into action and embodying the core tenets of shared humanity, we cannot model it or teach it to others. Shared humanity is more than a concept—it is a lifestyle and a way of being. It is a source of life in this present time of divisiveness.

Dr. Pascal & Crystal Losambe

In an age of binary thinking, shared humanity offers a third narrative. Currently in the US, there is a continuous current of divisiveness, blame, and shame, keeping others rooted in unforgiveness and isolation. The shared humanity narrative emphasizes interdependence, self-awareness, humility, respectful dialogue, the richness and power of our differences, the power of human connection, empathy, and compassion with action. The time is now to carry out this narrative boldly and call ourselves and each other to a higher standard, to set the temperature and not reflect it.

We hope that you will use this information to become more self-aware of your biases and avoid assumptions while striving for a deeper processing and understanding of each person. Additionally, we wish for readers to take what they have learned and strive towards acceptance and respect for others—including empathy and compassion. We also hope to foster resilience, inner confidence, and acceptance of self with the knowledge there is room for growth at every level. Finally, understand that we all want the same things for ourselves and our loved ones at the core of our being. We all have the same need for belonging, safety, and acceptance. Remember, when we are busy filling these needs, we cannot invest in others, be more productive and successful in our careers and organizations, and invest in our community.

The dream for this book and our company came from our vision for the future where tumultuous relationships are reconciled within organizations and social groups. We are hoping to lead the charge towards a world where:

- Communities come together across social barriers, emotional walls are torn down, and people forgive each other—moving forward together in unity.

- Someone can walk into a community, an organization, or a gathering and *not* have to scan the walls or count how many people of a certain racial group are there before they feel comfortable.

- People's first thoughts are not to question their safety or acceptance.

- Rejection and isolation are not more common than acceptance and positivity.

- People are open to learning from others, and if you do not understand, you still respond from a place of compassion, empathy, and respect.

- People can have constructive and productive dialogue and positive conversations about their differences. People can agree to disagree without inflicting harm or disrespect.

- Diversity and individual differences are valued, celebrated, and appreciated.

- Diversity is considered an asset and is utilized to bring success to organizations and communities.

- People are given the same benefit of the doubt and not judged prematurely based on preconceived notions about people groups.

Our company, Synergy Consulting Company desires to meet these needs in our society; the unmet needs have dramatically affected people's emotional, mental, and physical health. Much pain, stress, and anguish are caused by relational and social conflict, rejection, and divisiveness in groups, organizations, and communities. This has led to a decline in mental health and can even be translated into physical ailments, sickness, and disease.

Our desire is to:

- Provide a ministry of emotional and relational healing and reconciliation and to foster unity in organizations and communities.

- To empower and equip people and organizations with the necessary skills to create a culture of inclusion and acceptance of others for the greater good of our society, nation, and world.

- Affect our generation and future generations by leaving a legacy of peace, compassion, and unity that can continue beyond our natural lives.

Being Solution Minded

Some would argue that laws and policies have been passed to counteract historical narratives around diversity and equity. But, have these laws solved the problems? Do they make others feel like they belong, or do they create more division in society?

It is important to note that while laws and policies have been established to counter the narratives of the past, people are still sensitive to the persistent attitudes and mindsets that may impact an individual's sense of value and worth, especially for people of color. We cannot forget that people are behind every law, policy, practice, theory, or framework. It is the heart and motives of the person that matter. We cannot lose or forget our humanity.

The contextual memory people pull from is mainly incomplete due to societal narratives. For example, we refer to how little we hear about African American people's resilience, culture, and joy. Mattis and colleagues (2016) suggest for people who have endured centuries of dehumanization, discrimination, marginalization, and mistreatment society does not often paint the whole picture.[304]

After emancipation, people of African descent were exposed to dehumanizing conditions, poverty, residential segregation, economic exploitation, the absence of civil rights, and marginalization.

Yet, now the US Census Bureau reports:

- The life expectancy for African Americans is on the rise.[305]
- The number of African Americans living below the poverty line is declining.[306]
- The number of African Americans enrolling in college and universities, graduating, and earning professional degrees is rising.[307]

These above facts go largely unreported. Why? Perhaps negative social narratives at work?

When we align ourselves with absolute truths and frameworks, we often lose the humanity of people. We need each other in this narrative of unity and reconciliation. Conversely, individuals must acknowledge how some

---

[304] Mattis, J. S., Simpson, N. G., Powell, W., Anderson, R. E., Kimbro, L. R., & Mattis, J. H. (2016). "Positive psychology in African Americans." In E. C. Chang, C. A. Downey, J. K. Hirsch, & N. J. Lin (Eds.), *Positive psychology in racial and ethnic groups: Theory, research, and practice* (pp. 83–107). American Psychological Association. https://doi.org/10.1037/14799-005.

[305] Medina, L., Sabo, S., & Vespa, J. (2020). *Living longer: historical and projected life expectancy in the United States, 1960 – 2060.* United States Census Bureau. https://www.census.gov/content/dam/Census/library/publications/2020/demo/p25-1145.pdf.

[306] Creamer, J. (2020). *Inequalities persist despite decline in poverty for all major race and Hispanic origin groups.* United States Census Bureau. https://www.census.gov/library/stories/2020/09/poverty-rates-for-blacks-and-hispanics-reached-historic-lows-in-2019.html.

[307] US Census Bureau (2022). *Black education on the rise.* United States Census Bureau. https://www.census.gov/content/dam/Census/library/visualizations/2022/comm/black-education.pdf.

White people risked their lives to introduce a narrative of humanity (allyship). White individuals have stood and still stand in the gap and have worked to create an inclusive society. These are essential narratives to amplify.

We have spoken to many individuals about the deficit of positive narratives of African Americans. Often people feel uneasy when we speak with them about this because they infer that we should be happy with where we are. However, we mean quite the opposite. Progress can be measured according to how far we have come compared to the past *and* where we are compared to where we want to be. One does not cancel out the other. Acknowledging the past and revisiting the triumphs and struggles allows us to become more solution-minded in moving towards the goal. Being stuck in perpetual deficit narratives will keep us stuck, but engaging towards solutions with hope, courage, and interdependence will enable us to progress. We all belong in this conversation around race because it impacts us all. We all need something from this narrative, so we must humanize one another.

Who Holds Responsibility

Our shared humanity is the entry point into these conversations. Many people, regardless of their religion, gender, race or ethnicity, ability, and so on, empathize and connect with concepts in this book, like the idea of high-context dependency. Most have been in new environments where they feel uncertain or like they do not belong. This is not just a minority experience—it is part of our shared humanity.

Yet, our stories are not horizontal or parallel. For example, research indicates that racial and ethnic minorities experience microaggressions, negative bias, stereotype threat, imposter syndrome, and unbelonging—triggers of high context dependency—at a much higher rate. People with marginalized identities are likely to enter the high-context-dependent state more often than other identity groups. However, we can engage with each other from that understanding. It is all about finding the common thread to start the conversation and seeing one another as human. The fantastic thing about the brain is that when someone tells a story or relays a unique experience, humans can internalize the information and feel similar emotions.

The Present and Future

The concept of shared humanity is founded in the fundamental precept of Ubuntu, which as we explained at the beginning of the book means that "my humanity is caught up [and is] inextricably bound up in yours." In South Africa, where Pascal spent most of his formative years, specifically in the Xhosa culture, the word Ubuntu is used as a greeting and is translated as "I am because we are." This greeting reminds us that we have a shared responsibility to create environments of belonging where we and others can flourish with every human interaction.

Living according to the foundational tenets of Ubuntu and embracing our shared humanity requires many of us to go against our yearnings for comfort, the default way we build community—that is, forming connections with people who have similar identities, beliefs, values, and worldviews—and our natural tendency to create "us versus them" categorizations.

Shared humanity calls for us to live by the platinum rule—treat others the way they want to be treated. The platinum rule underscores the need for authentic relationships, an open and humble heart in listening, truly hearing what others are going through, and the desire to know how to support them. If history reveals a core truth to us, interdependence is needed to achieve the most significant potential in our schools, an arena where most people are at least somewhat invested.

Our shared humanity begins with the understanding that every behavior is motivated by a desire to fulfill specific human needs. Even though our needs are universal, individuals have different experiences, perspectives, contexts, and truths that affect them and motivate them to fulfill these needs in different ways.

When we begin to view ourselves and others from the lens of our shared humanity, we will feel compelled to practice self-awareness through reflection and introspection and begin to understand one another's needs, desires, impulses, thoughts, and emotions. Consequently, individuals will hold themselves accountable for their thoughts and actions and treat each other with respect and dignity.

Our shared humanity requires humility and an acknowledgment of the uniqueness of others. Moreover, it is a call for individuals *not* to lead with assumptions and preconceived notions of who they believe other people to be.

While it is essential to recognize an individual's uniqueness and individuality, socially constructed categories like race, ethnicity, gender, ability, sexual orientation, and class cannot be ignored. These categories affect an individual's sense of self and how others see them. These elements matter because they are part of individual identity and experience. Therefore, we must be wary of both the biases we have towards others and the inter-group biases that have been perpetuated by society.

Through respectful dialogue, people can understand one another's backgrounds and experiences, which make up who they are. Thus, people can build empathy for those around them, even if they do not share or fully understand where others are coming from. Individuals will also know that every person's desires lead back to our universal needs.

It is important to emphasize that shared humanity relies not only on individuals but also on leaders and organizations creating systems and processes to ensure that the foundational elements of shared humanity exist and persist within the organizations. The benefit of having systems and procedures in place is so that this culture can outlive any individual. Those who enter an organization will have a sense of belonging and also be aware of the behavioral expectations of the organization.

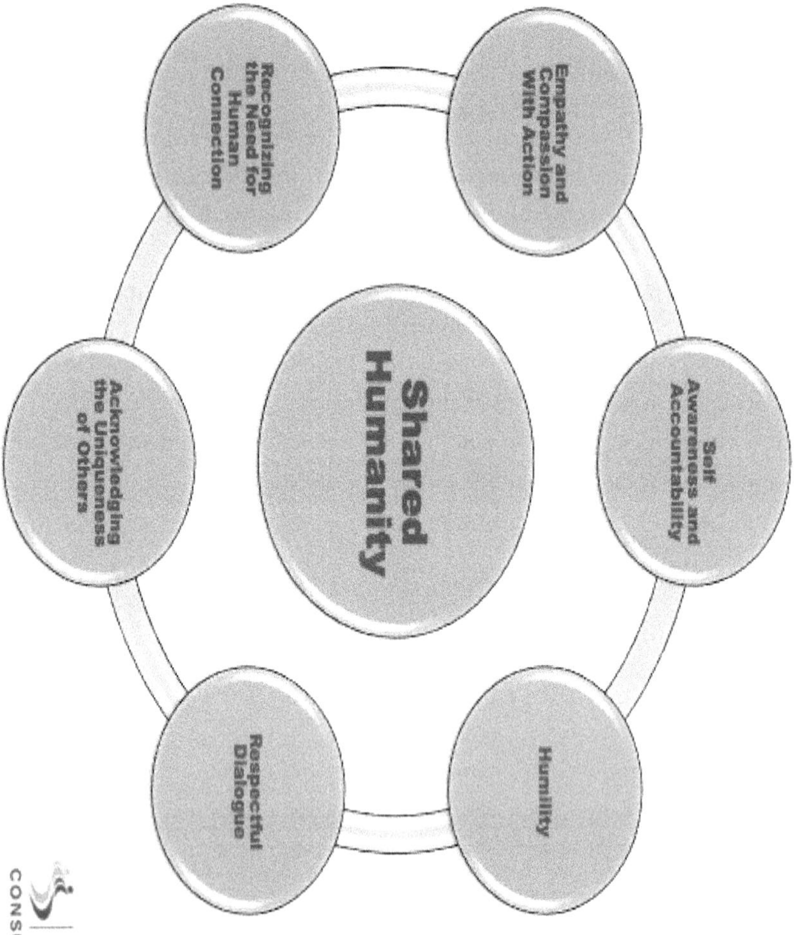
Shared Humanity
Recognizing the Need for Human Connection
Empathy and Compassion With Action
Self Awareness and Accountability
Humility
Respectful Dialogue
Acknowledging the Uniqueness of Others

SYNERGY CONSULTING COMPANY

The Power of Vision

Throughout our writing of this book, there were moments when we wavered. Amid anti-Asian sentiment, the deaths of African American individuals, and a "cancel culture" that silenced the majority, we began to question our goals. But as we discussed our trepidation with a friend, he asked us a simple question, "What do you believe? Not intellectually, but in your heart. What do you believe in your heart?"

So, we will ask you: What do you believe in your heart now that you have read this book? Do you see things differently? Are you more aware of your words and actions? Are you motivated to be an ally for those who need it? Are you ready to embrace change?

If you answered yes to the last four questions, keep in mind the future state you are trying to achieve, as these behaviors may be new to you or may cause you to step out of your comfort zone. However, the power of vision cannot be overemphasized. When you visualize a desired future state, your brain cannot distinguish between fact and fiction. Hence, it believes what you visualize. It allows you to remember that vision. Being intentional about continually visualizing your desired future state is important for encoding—that is, for your brain to hold on to that vision (biological encoding).

Another way to perform biological encoding is by writing down your vision. When you write down your vision for who you want to be in this narrative around race and ethnicity, your brain visualizes that future state, and revisiting it will enable your brain to become more confident in acting it out. You can also trick your brain by visualizing your undesired state, and your brain will work to avoid it. When you have a vision, the brain does not rely on the past or its defaults but starts to develop a map for the future.

But before you start, remember a few key points.

Your mind has selective amnesia.

Your brain does not remember specifics. It remembers things in general. Many of you, if not all of you, have navigated change effectively at some point or another, and you likely have a general idea of how you successfully navigated a transition in the past. However, your brain only remembers things in generalities and forgets details. The connections to those past change events may be weaker, but as you sit and try to revisit those moments, you may find specific things that you did and specific characteristics that you demonstrated to embrace that change. It may be essential to remember the details of those situations where you successfully navigated change and those where you did not and learn from those situations.[308]

Make the change bigger than just you and create the path.

Remember Abraham Maslow's last need: self-transcendence. We need to live for something beyond ourselves to lead a purposeful life. So be clear about what you want to achieve and why you want to achieve it. What does this change mean for you and others? What does success look like along the way? Do you have short, mid-term wins? What happens when you fail? How often will you have meetings with yourself? Creating a path forward will allow you to focus on the ideal and steps you can take to achieve that ideal.[309]

The journey is more important than the destination: dopamine.

Dopamine is a neurochemical that is increased when we obtain rewards. It enables individuals to become more motivated to get a specific outcome. When we are on our journeys and fulfill our commitments, display courage in the face of discomfort, or become a confidant to someone who has been affected by some societal situation. Dopamine is released at that moment. Dopamine enables us to want to pursue that behavior again. In essence, pleasure is found on the journey, and we become more confident in pursuing certain behaviors because of dopamine.

In essence, if we focus too much on high level goals, such as the organization being more diverse, changing laws and policies, etc., and we do not see those goals realized, we may become less motivated. However, suppose we attach our sense of reward to individual responsibility, attitudes, behaviors, and actions. In that case, the likelihood is that we will release dopamine when we achieve things on the journey and stay motivated. Therefore, focusing on the journey can be more encouraging than focusing

---

[308] Gravitz, L. (2019). "The forgotten part of memory." *Nature*, *571*(7766): S12-S12.

[309] Koltko-Rivera, M. E. (2006). "Rediscovering the later version of Maslow's hierarchy of needs: Self-transcendence and opportunities for theory, research, and unification." *Review of General Psychology*, *10*(4): 302-317.

on the outcome. Furthermore, embracing discomforts will allow you to break defaults and learn from yourself.[310]

Equipped with the knowledge, skills, and tools you have gained from this book, the next focus is on "being," believing, and living out what we have learned with the skills and tools given. We hope you will continually reflect on your practice, enhance your learning, and hold yourself accountable for using those skills and tools. When shared humanity becomes your ethos, it will impact your thoughts and actions, and in the end, everyone triumphs together.

---

[310] Lembke, A. (2021). *Dopamine Nation: Finding Balance in the Age of Indulgence.* Dutton.

Appendix

Understanding the Brain

For those of you who are interested in a more detailed understanding of the science behind our need to belong, we have included this material on the brain. There are functions within our brains that promote the desire to belong, and therefore, it can help us understand this anatomical motivator, the human brain. Knowing how it functions can make a difference in our individual decisions as well as the organizational strategies we choose when we do DEI work. Therefore, it is a really critical piece of the puzzle.

The brain is an organ encased inside a sturdy skull and acts as the command center that controls all our bodily functions while making sense of the world around us. Most miraculously, the brain controls those functions outside our consciousness—those things we do automatically, such as heart rate, hunger pains, blood pressure, and pain responses. If we can see how these functions and responses are controlled by this amazing organ, we can comprehend the human need to belong at its most basic level.

The Structures of the Human Brain

The brain is composed of many parts. The so-called "lower structures" of the brain are the brainstem and cerebellum. Scholars nickname this part of the brain the "reptilian" or "lizard" brain because it resembles structures that exist in reptiles. It is also the oldest portion of the brain. This is the first part of the brain to develop and mature in the womb. This part of the brain controls states such as focus, arousal, alertness, thirst, and hunger and modulates and regulates other physiological body states such as heart rate, blood pressure, breathing, and temperature. This part of the brain is first to develop and is always "on" since it is essential for survival.[311, 312]

A network of cells is found in the brainstem called the **reticular activating system (RAS)** which controls alertness and attention. The RAS filters out unimportant information but pays close attention to information that is important to an individual.[313] For example, for parents, imagine your child starts crying as you are falling asleep. Would you suddenly become alert? Yes, you would. This is because your RAS is sensitive to your child's cry. The RAS is sensitive to your environment and is always in scan mode. It is sensitive to things like novelties, things that you find vital, as well as signs of belongingness and unbelonging. It is perceptive to the presence and absence of things, behaviors, and actions affirming and validating you. If you remember, the brainstem is one of the structures that is first to develop, so the RAS is involved in paying attention to smells, sounds, faces, and so on from a young age.

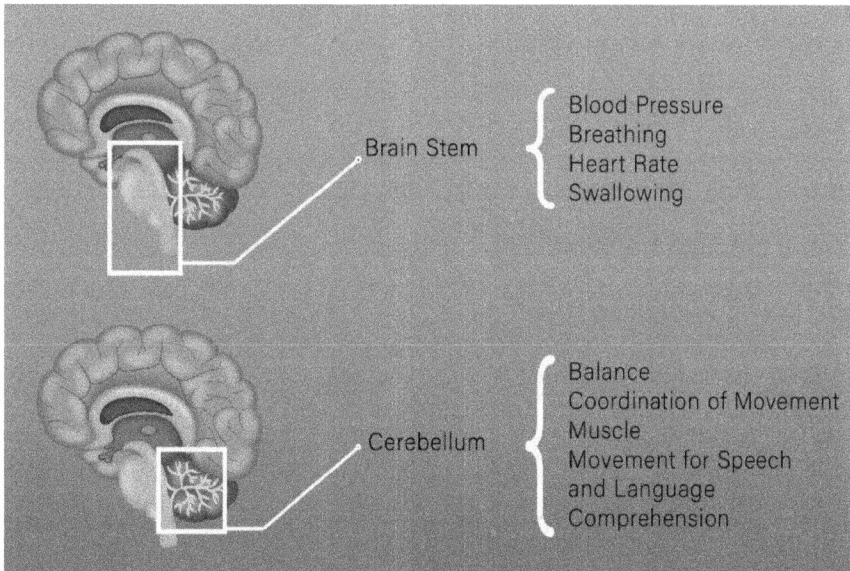

Brain Stem
{ Blood Pressure
Breathing
Heart Rate
Swallowing

Cerebellum
{ Balance
Coordination of Movement
Muscle
Movement for Speech
and Language
Comprehension

---

[311] Siegel, D. J. (1999). *The Developing Mind: Toward a Neurobiology of Interpersonal Experience.* Guilford Press.

[312] Hammond, Z. (2014). *Culturally Responsive Teaching and the Brain: Promoting Authentic Engagement and Rigor Among Culturally and Linguistically Diverse Students.* Corwin Press.

[313] Garcia-Rill, E. (2015). *Waking and the Reticular Activating System in Health and Disease.* Academic Press.

Figure: Illustrates parts of the primitive brain (the brainstem and cerebellum)

The Thalamus and Senses

Take a moment and think about the location of the brain. It is encased in the skull devoid of contact from the outside world, yet it controls all the actions we perform. However, for the brain to perform its functions, it relies on all of our senses: taste, touch, hearing, smell, and sight. The top of the brain stem is where those senses get interpreted through the brain structure known as the thalamus. The thalamus receives information from our senses, except for smell, and is connected to many other portions of the brain that enable the brain to process the information we receive, mainly from the outside world.[314, 315, 316]

The Limbic System

The next part of the brain is the limbic system, which is essential to supporting our emotions, motivation, long and short-term memory, and goal-directed behavior. The limbic system sits right above the brainstem (or reptilian region) and plays an important role in the coordination of the lower and higher structures of the brain (more on that later). In addition, the limbic system includes essential structures like the hippocampus, hypothalamus (and the pituitary gland), and amygdala. Let us look into these structures in more detail:

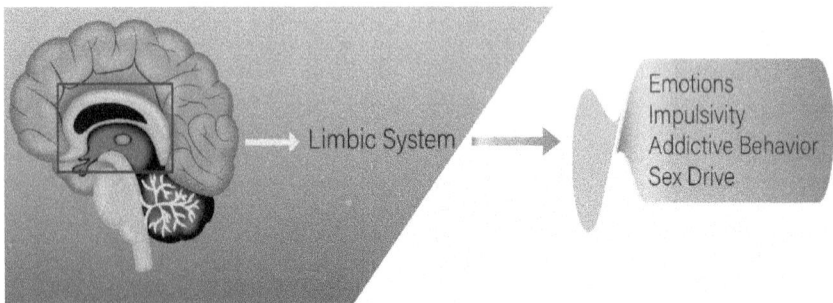

---

[314] Wolfe, P. (2010). *Brain Matters: Translating Research into Classroom Practice.* ASCD.

[315] Hammond, Z. (2014). *Culturally Responsive Teaching and the Brain: Promoting Authentic Engagement and Rigor Among Culturally and Linguistically Diverse Students.* Corwin Press.

[316] Bard, A., & Bard, M. G. (2002). *The Complete Idiot's Guide to Understanding the Brain.* Penguin.

Hippocampus

Memories are bits of information gathered through our lived experiences, stored, and retrieved. For example, when we learn or experience something, our brains form circuits that can be accessed when recalling a behavior or action. The hippocampus is primarily involved in that memory formation, particularly short-term memory formation, and is later housed in the hippocampus and other regions. So how does short-term memory differ from long-term memory? Short-term memory holds memories for about 30 seconds. For example, you may have forgotten what you read at the top of this page a few minutes ago but remember what you read less than a minute ago. If you are like many, when trying to remember a phone number, you keep repeating it as you type it into your phone. Once it is typed in, the number vanishes from your consciousness. That is an example of you using your short-term memory.

We also need to distinguish short-term memory from working memory. Working memory is when you retain or retrieve small amounts of information and use that information towards a goal or objective. Working memory can be temporarily accessed from the long-term memory to be used in a story or can be newly learned information that has some purpose. For example, you may see someone who looks like your third-grade teacher. Then you recall the face of your third-grade teacher and that of the person you are looking at and hold onto the information long enough to tell a friend—your working memory at play.

Last, long-term memories are the bits of information stored in your brain for months to several years. It is like saving a file on the hard drive of your computer. Often, things such as values, beliefs, events, and occasions have either been reinforced throughout your life or coupled with heightened emotion, like an anniversary, significant achievement, birthday, and so on. These are stored in your long-term memory. However, mundane everyday experiences are likely not stored in long-term memory. For example, if I ask you what you ate for breakfast on a Tuesday last month, you probably do not recall it because it has been shed by the brain as insignificant.[317, 318, 319]

Amygdala

The limbic system also contains an almond-shaped, fingernail-sized structure called the amygdala. The amygdala is directly connected to the thalamus, hypothalamus, pituitary, and brainstem. It plays a vital role in initiating specific emotional reactions in response to events. Many professionals term the amygdala as the fear center or danger detector. In fact, it plays a role in affecting facial expressions and protective postures, like crouching. It is intricately involved in regulating the fight, flight, or freeze response through its influence on releasing hormones such as adrenaline, noradrenaline, and cortisol. Its connection to the hypothalamus initiates physiological changes such as increased blood pressure, heart rate, muscle contractions, respiration, and perspiration.

Again, the amygdala and the hippocampus are connected. The hippocampus plays a vital role in retrieving memories associated with different environments, circumstances, or situations. For example, suppose you continually experience unbelonging in an environment and then enter it. In that case, the hippocampus may recall those memories and indicate to the amygdala that it needs to prepare for an unpleasant experience.

Researchers trying to understand the function of the amygdala found that individuals who lack a large portion of their amygdala due to seizures, disease, or injury do not recognize expressions or body language associated with anger, screams, scowls, and angry voices. Moreover, their fear conditioning, learning what to fear, is also compromised from situation to situation.

The Neocortex

We have discussed structures associated with the reptilian brain such as the brainstem and the limbic system. These areas are essential to forming memories and the unconscious regulation of critical physiological processes.

[317] Wolfe, P. (2010). *Brain Matters: Translating Research into Classroom Practice.* ASCD.

[318] Hammond, Z. (2014). *Culturally Responsive Teaching and the Brain: Promoting Authentic Engagement and Rigor Among Culturally and Linguistically Diverse Students.* Corwin Press.

[319] Bard, A., & Bard, M. G. (2002). *The Complete Idiot's Guide to Understanding the Brain.* Penguin.

This section will address the part of the brain that controls our conscious thoughts, like processing and talking about your emotions and being consciously aware of your thoughts.

When you look at a picture of the brain, you will likely be looking at the most significant portion of the brain called the cerebrum. The cerebrum is covered by a thin layer known as the cerebral cortex or the neocortex. The word cortex is derived from the Latin word "bark," and it is appropriately named because it covers the brain's outer layer. When we think of bark, we often think of a dark color. The cortex is also called gray matter due to its darker color.

Different areas of the cortex are associated with distinct functions. To generally describe these areas, scientists describe four regions, also known as lobes. They are:

1. The frontal lobe, located at the front of the brain;
2. The parietal lobe, located in the middle of the brain;
3. The temporal lobe, located on the lower side of the brain, right near the ears; and
4. The occipital lobe, located at the back of the head.

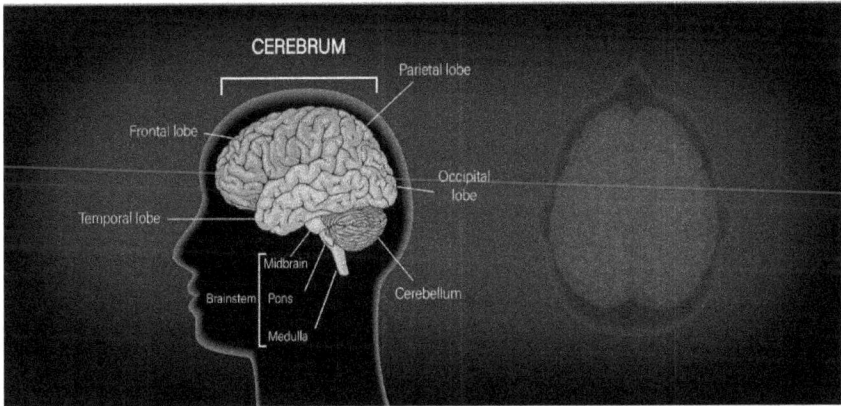

The occipital lobe, also called the visual cortex, is primarily responsible for visual processing information in the region of the brain. The temporal lobes function in processing auditory stimuli, the parietal lobe in spatial awareness and movement, sensations of pressure and pain, and other general tactile information from the environment. The frontal lobe controls problem-solving, processing of emotions, and complex thoughts.

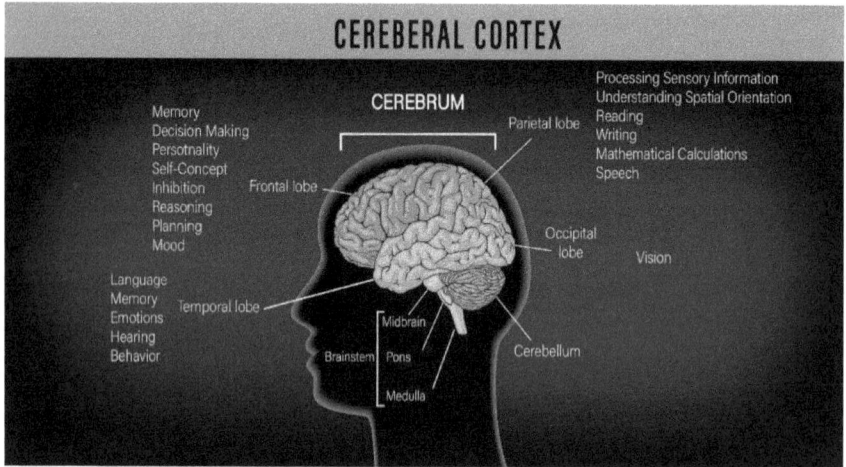

Figure: This image illustrates the different regions (or lobes) of the brain.

The posterior portion of the frontal lobe also contains the brain's motor cortex, which can initiate and direct muscular movement. In fact, every part of our body is controlled by a specific region in the motor cortex.

There is an area at the front of the brain, called the prefrontal cortex, worth mentioning. The prefrontal cortex is more prominent in humans than other species, separating humans from them. This part of the brain takes information from both inside and outside of you and makes sense of all of it. For example, associations between spoken words and recalled objects are created. Moreover, the bold statement made earlier, where we explained that the prefrontal cortex is what separates us from other species, means we have the flexibility of making sense of our worlds, adapting to different environments, understanding and visualizing someone's past, present, and future, and making sense of dreams.[320, 321, 322, 323]

Neurons

We said earlier that the brain is an organ. In what is known as the vertical organization of life, cells make up tissues, and tissues make up organs. Therefore, the most essential part of the brain is a cell called a neuron. There are about 100 billion neurons in the brain that are connected to one another and communicate through electrical and chemical signals. Neurons are so small that researchers say that 30,000 neurons can fit on the head of a pin.

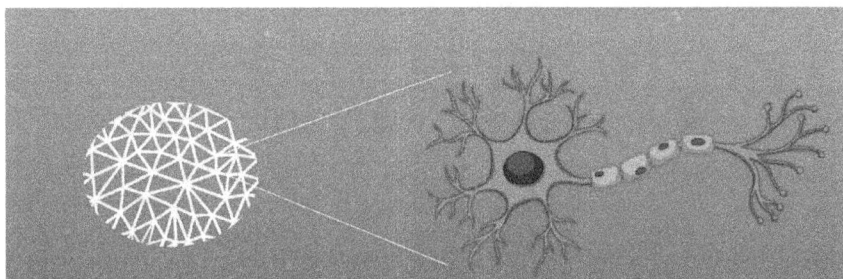

Figure: Brain cells (neurons) and also depicts an individual brain cell

A neuron contains a cell body that houses the nucleus, which includes the cell's genetic material, short finger-like projections called dendrites (from the word *dendra*, Greek for "tree"), and a long string-like structure called an axon, which is covered by a fatty substance called myelin. Think of myelin like insulation around a phone line.

---

[320] Carey, J. (1990). *Brain Facts: A Primer on the Brain and Nervous System.*

[321] Wolfe, P. (2010). *Brain Matters: Translating Research into Classroom Practice.* ASCD.

[322] Hammond, Z. (2014). *Culturally Responsive Teaching and the Brain: Promoting Authentic Engagement and Rigor Among Culturally and Linguistically Diverse Students.* Corwin Press.

[323] Bard, A., & Bard, M. G. (2002). *The Complete Idiot's Guide to Understanding the Brain.* Penguin.

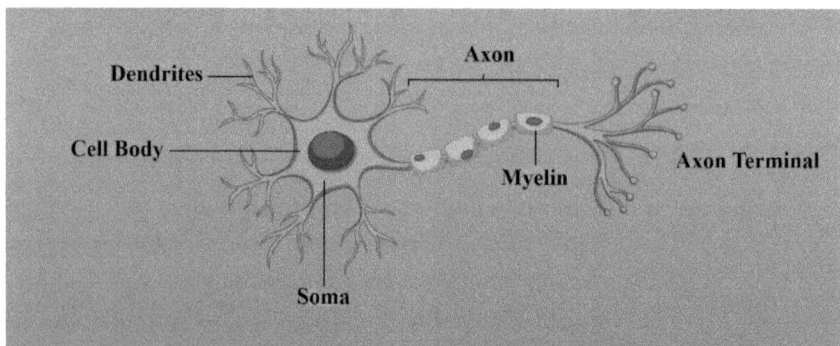

Figure: The parts of a neuron (dendrites, cell body, axon, myelin)

When we learn or experience something, neurons connect and communicate with one another. Neurons do not physically connect with one another but leave a small space between each other called a synapse. The synapse allows one neuron to pass on electrical and chemical signals to the next. For example, imagine you are at a dance and told to find a partner. You are instructed to stand face-to-face but not touch. The dancers are the connecting neurons. The space between them represents the synapse. That is how neuronal connections work. They find their partners without touching and communicate electrically and chemically. The axon of one neuron will connect with the dendrites of the next neuron.

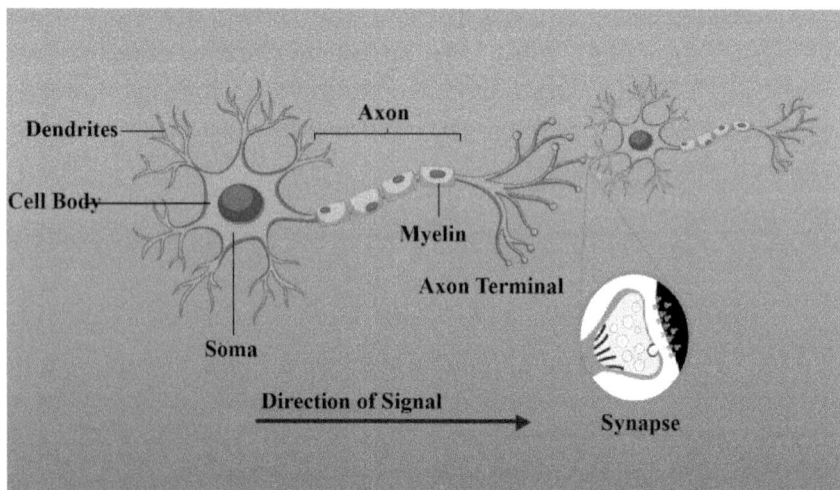

Figure: Neurons "connected" to one another and the formation of a synapse

Why is this important? When we learn or experience something new, neurons connect and create a neural pathway to a particular brain region where the signal or stimuli will be processed. For example, imagine reading an article and coming across a word you do not understand. You do not recognize or understand the word because you have not encountered it before. That is, there are no existing neuronal connections. So, a new pathway must be created with the existing neurons in the brain:

- Some neurons in the occipital lobe, specifically the visual cortex, form routes to recognize how the word is spelled.

- The part that processes auditory stimuli in your temporal lobe is organized when you pronounce the word.

- Neurons in your neocortex are organized to associate the word with preexisting knowledge when you look up its definition.

Now, the more you encounter the word and practice using it, the stronger those connections become and the more likely you are to remember the word. Suppose you are reading a magazine and encountering the same word again. The neural circuits you created when learning the word start to activate, and you remember the meaning. In general, all the memories you make—your values, beliefs, events, emotions, words, and images—have neuronal circuits in the brain that are recalled and accessed when the context allows. The pathways that are not reinforced are pruned away, and those that are, remain.[324, 325, 326, 327]

The Nervous System

The nervous system consists of the brain and spinal cord or the central nervous system and nerves or the peripheral nervous system. Even though the previous sections focused on the brain, your body comprises a highway system, more like a network of cables called neurons. They exist all over the body in bundles we call nerves that contact every part of our body. Yes, even

---

[324] Carey, J. (1990). *Brain Facts: A Primer on the Brain and Nervous System.*

[325] Wolfe, P. (2010). *Brain Matters: Translating Research Into Classroom Practice.* ASCD.

[326] Hammond, Z. (2014). *Culturally Responsive Teaching and the Brain: Promoting Authentic Engagement and Rigor Among Culturally and Linguistically Diverse Students.* Corwin Press.

[327] Bard, A., & Bard, M. G. (2002). *The Complete Idiot's Guide to Understanding the Brain.* Penguin.

your heart, lungs, fingertips, and toes, for example. These nerves are connected directly or indirectly to the brain and spinal cord, making sense of internal and external stimuli. Different parts of your brain make sense of the stimuli and activate your body to act on them.

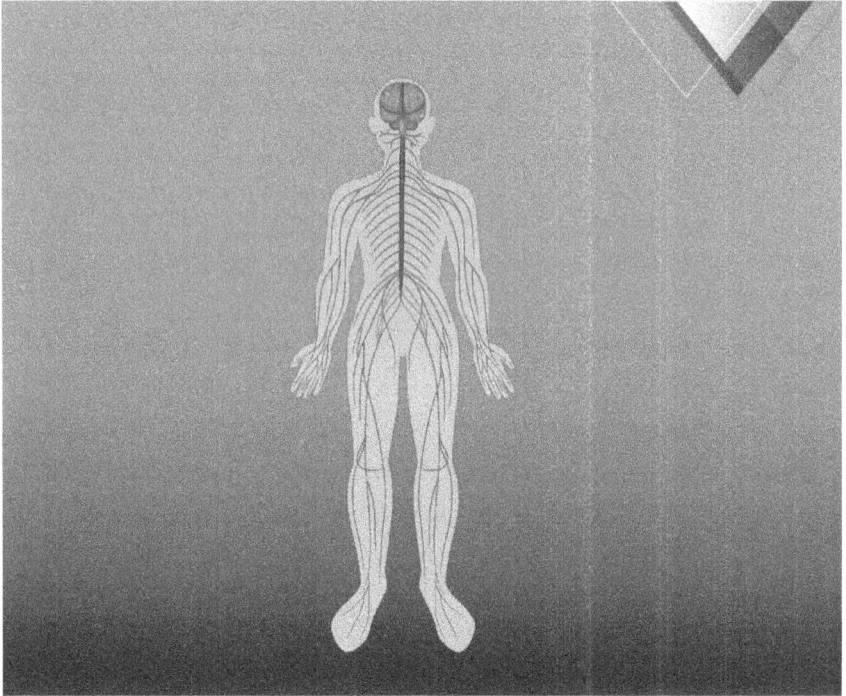

Figure: Bundle of nerves that exist throughout the body

There are three types of nerves—sensory, connector, and motor nerves. Sensory nerves carry information to the brain from sense organs, such as the nose, ears, eyes, skin, tongue (sensory input). As the neurons in the brain make sense of the signal (integration), decisions are made. The signal is passed onto connector neurons found in the brain and the spinal cord, which connects sensory and motor neurons, and the motor neurons take messages from the brain and spinal cord and send it to the muscles or glands that release chemicals in the body that regulate bodily functions. It is important to note that some parts of the nervous system are controlled by conscious thoughts such as, "I need to lift my hand," and unconscious processes, like a heartbeat.

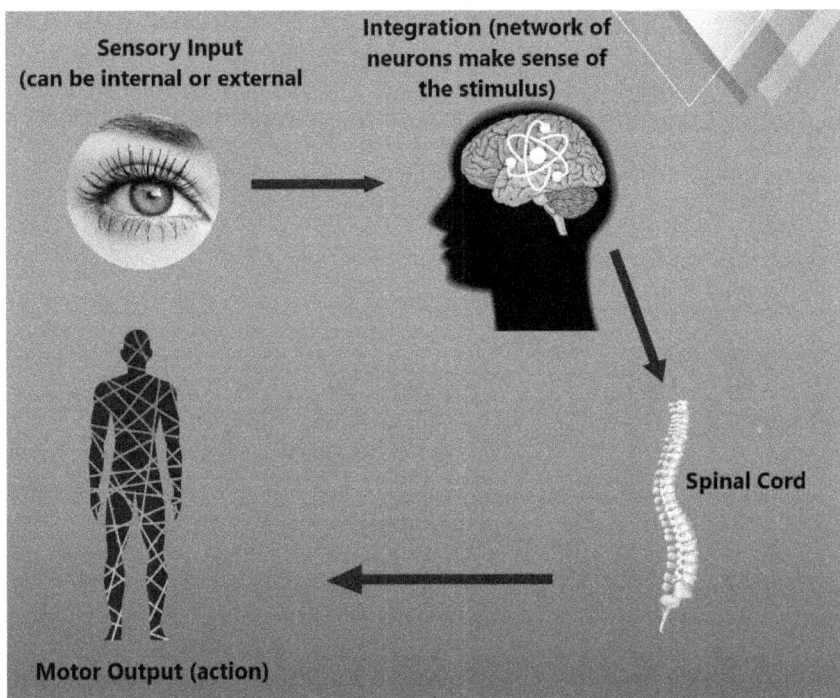

Figure: Sensory, integrative, and motor functions of the nervous system

The Mind-Brain Connection

People use the terms brain and mind interchangeably. However, the brain and mind are distinct entities. Using a computer as an analogy, the brain is the hardware, and the mind is the software. Much like the computer's screen, chips, and hard drive, the brain contains many cells and proteins that stabilize and allow it to function. Like a computer, the brain is the structure you can see, touch, and hold. The software, what you cannot see, interprets the information and data made available to the hardware. In much the same way, the mind allows us to make sense of the data it receives. The brain cannot exist without the mind, and the mind cannot exist without the brain. They are intricately connected.

The mind governs a human being and its interactions with its environment. It consists of the human soul, a person's values, beliefs, thoughts, and character, the intellect, and the subjective world perception. There are three main aspects: the conscious mind, the preconscious mind, and the unconscious

261

mind. To illustrate these essential parts, visualize an iceberg. Then read the following descriptions below:

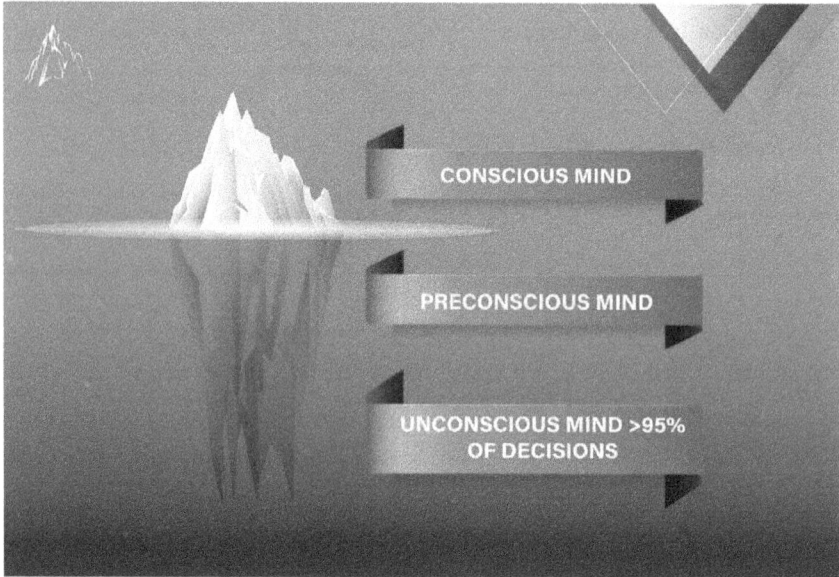

Conscious

The visible part of the iceberg above the water is like the conscious mind. These are things a person is aware of at any given moment: thoughts, feelings, memories, and emotions. Individuals can easily talk about their conscious mind's thoughts, feelings, and emotions because they feel them in real-time. For example, if you say to yourself, "I'm feeling hungry," your conscious mind is aware of your hunger and thinking of ways to get food.

Preconscious

The portion of the iceberg just beneath the surface is called the preconscious mind. This is often referred to as available memory. For example, if you were asked to give someone's phone number or address, that information may easily be accessible. Even though you were not consciously thinking about the phone number, you can easily recall it. When you remember a phone number or address, your preconscious thoughts move to the conscious mind.

Unconscious

The iceberg's deepest parts are analogous to the unconscious mind. This is the storehouse of all memories, motives, values, worldviews, and beliefs. Here is the thing—the unconscious mind usually dictates a person's reactions and helps individuals make sense of their world. The unconscious mind stores pictures and associates those pictures with a negative or positive belief.[328, 329, 330, 331]

Summary

Understanding science about the brain is essential when actively creating social change. Again, what is the best course of correction or enhancement to address any issue or concern? You need to address the root cause. The mind and brain are the core processor of your feelings and beliefs. Those feelings and beliefs then impact how we react in society.

For example, suppose you are walking down the street, and a dog is coming toward you. You know nothing about the dog or its owner, but you had a negative experience with a dog as a child. Your hands begin to sweat, and you freeze, trying to decide to cross the street or not. What is happening? Your brain is processing a flight, fight, or freeze response to stimuli with a negative memory association. Is this good or bad? Well, that all depends on whether you want to grow and get over your fear.

The beauty of the mind and brain is that they can be reprogrammed. For example, if you take the scenario above, how could you alter your fear response to seeing a dog? Well, over time, with more and more exposure to positive interactions with dogs, your brain will create new positive memories

---

[328] Boag S. "Conscious, Preconscious, and Unconscious." *Encyclopedia of Personality and Individual Differences.*

[329] Cherry, K. (2020, December 9). *The Structure and Levels of the Mind According to Freud.* Verywell Mind. January 24, 2022. https://www.verywellmind.com/the-conscious-and-unconscious-mind-2795946.

[330] Walinga, J. (2014, October 17). *2.2 "Psychodynamic Psychology." Introduction to Psychology,* 1st Canadian Edition. January 24, 2022. https://opentextbc.ca/introductiontopsychology/chapter/2-2-psychodynamic-and-behavioural-psychology/.

[331] Mcleod, S. (2015, January 1). "Freud and the Unconscious Mind." *Unconscious Mind: Simply Psychology.* January 24, 2022. https://www.simplypsychology.org/unconscious-mind.html.

so that when you see a dog walking toward you, you will no longer be paralyzed with fear.

About Kharis Publishing:

Kharis Publishing, an imprint of Kharis Media LLC, is a leading Christian and inspirational book publisher based in Aurora, Chicago metropolitan area, Illinois. Kharis' dual mission is to give voice to under-represented writers (including women and first-time authors) and equip orphans in developing countries with literacy tools. That is why, for each book sold, the publisher channels some of the proceeds into providing books and computers for orphanages in developing countries so that these kids may learn to read, dream, and grow. For a limited time, Kharis Publishing is accepting unsolicited queries for nonfiction (Christian, self-help, memoirs, business, health and wellness) from qualified leaders, professionals, pastors, and ministers. Learn more at: About Us - Kharis Publishing - Accepting Manuscript